THE
DEEP LEARNING
WITH PYTORCH
WORKSHOP

Build deep neural networks and artificial
intelligence applications with PyTorch

Hyatt Saleh

THE DEEP LEARNING WITH PYTORCH WORKSHOP

Author: Hyatt Saleh

Reviewers: Tim Hoolihan, Narinder Kaur Saini, Anuj Shah, Nahar Singh, and Subhash Sundaravadivelu

Managing Editor: Anush Kumar Mehalavarunan

Acquisitions Editors: Royluis Rodrigues, Kunal Sawant, Sneha Shinde, Anindya Sil, and Karan Wadekar

Production Editor: Shantanu Zagade

Editorial Board: Megan Carlisle, Samuel Christa, Mahesh Dhyani, Heather Gopsill, Manasa Kumar, Alex Mazonowicz, Monesh Mirpuri, Bridget Neale, Dominic Pereira, Shiny Poojary, Abhishek Rane, Brendan Rodrigues, Erol Staveley, Ankita Thakur, Nitesh Thakur, and Jonathan Wray

First published: July 2020

Production reference: 2230221

ISBN: 978-1-83898-921-7

Published by Packt Publishing Ltd.

Livery Place, 35 Livery Street

Birmingham B3 2PB, UK

WHY LEARN WITH A PACKT WORKSHOP?

LEARN BY DOING

Packt Workshops are built around the idea that the best way to learn something new is by getting hands-on experience. We know that learning a language or technology isn't just an academic pursuit. It's a journey towards the effective use of a new tool—whether that's to kickstart your career, automate repetitive tasks, or just build some cool stuff.

That's why Workshops are designed to get you writing code from the very beginning. You'll start fairly small—learning how to implement some basic functionality—but once you've completed that, you'll have the confidence and understanding to move onto something slightly more advanced.

As you work through each chapter, you'll build your understanding in a coherent, logical way, adding new skills to your toolkit and working on increasingly complex and challenging problems.

CONTEXT IS KEY

All new concepts are introduced in the context of realistic use-cases, and then demonstrated practically with guided exercises. At the end of each chapter, you'll find an activity that challenges you to draw together what you've learned and apply your new skills to solve a problem or build something new.

We believe this is the most effective way of building your understanding and confidence. Experiencing real applications of the code will help you get used to the syntax and see how the tools and techniques are applied in real projects.

BUILD REAL-WORLD UNDERSTANDING

Of course, you do need some theory. But unlike many tutorials, which force you to wade through pages and pages of dry technical explanations and assume too much prior knowledge, Workshops only tell you what you actually need to know to be able to get started making things. Explanations are clear, simple, and to-the-point. So you don't need to worry about how everything works under the hood; you can just get on and use it.

Written by industry professionals, you'll see how concepts are relevant to real-world work, helping to get you beyond "Hello, world!" and build relevant, productive skills. Whether you're studying web development, data science, or a core programming language, you'll start to think like a problem solver and build your understanding and confidence through contextual, targeted practice.

ENJOY THE JOURNEY

Learning something new is a journey from where you are now to where you want to be, and this Workshop is just a vehicle to get you there. We hope that you find it to be a productive and enjoyable learning experience.

Packt has a wide range of different Workshops available, covering the following topic areas:

- Programming languages

- Web development

- Data science, machine learning, and artificial intelligence

- Containers

Once you've worked your way through this Workshop, why not continue your journey with another? You can find the full range online at http://packt.live/2MNkuyl.

If you could leave us a review while you're there, that would be great. We value all feedback. It helps us to continually improve and make better books for our readers, and also helps prospective customers make an informed decision about their purchase.

Thank you,
The Packt Workshop Team

Table of Contents

Preface i

Chapter 1: Introduction to Deep Learning and PyTorch   1

Introduction 2

Why Deep Learning? 2

   Applications of Deep Learning 4

Introduction to PyTorch 5

   GPUs in PyTorch 6

   What Are Tensors? 7

   Exercise 1.01: Creating Tensors of Different Ranks Using PyTorch 9

   Advantages of Using PyTorch 10

   Disadvantages of Using PyTorch 11

   Key Elements of PyTorch 12

      The PyTorch autograd Library...12

      The PyTorch nn Module 13

   Exercise 1.02: Defining a Single-Layer Architecture 15

      The PyTorch optim Package 17

   Exercise 1.03: Training a Neural Network 19

   Activity 1.01: Creating a Single-Layer Neural Network 21

Summary 23

Chapter 2: Building Blocks of Neural Networks   25

Introduction 26

Introduction to Neural Networks 27

What Are Neural Networks? 28

Exercise 2.01: Performing the Calculations of a Perceptron 29

Multi-Layer Perceptron 30

The Learning Process of a Neural Network 31

    Forward Propagation 32

    The Calculation of Loss Functions 36

    Backward Propagation...37

    Gradient Descent..39

Advantages and Disadvantages 40

    Advantages...40

    Disadvantages...40

Introduction to Artificial Neural Networks 42

Introduction to Convolutional Neural Networks 43

Introduction to Recurrent Neural Networks 48

Data Preparation 50

Dealing with Messy Data 50

Exercise 2.02: Dealing with Messy Data 52

Data Rescaling 57

Exercise 2.03: Rescaling Data 58

Splitting the Data 59

Exercise 2.04: Splitting a Dataset 60

Disadvantages of Failing to Prepare Your Data 63

Activity 2.01: Performing Data Preparation 64

Building a Deep Neural Network 65

Exercise 2.05: Building a Deep Neural Network Using PyTorch 67

Activity 2.02: Developing a Deep Learning Solution
for a Regression Problem 70

Summary 71

Chapter 3: A Classification Problem Using DNN   73

Introduction 74

Problem Definition 74

    Deep Learning in Banking 75

    Exploring the Dataset 77

    Data Preparation 81

    Building the Model 88

    ANNs for Classification Tasks 89

    A Good Architecture 90

    PyTorch Custom Modules 91

    Exercise 3.01: Defining a Model's Architecture

    Using Custom Modules 93

    Defining the Loss Function and Training the Model 94

    Activity 3.01: Building an ANN 98

Dealing with an Underfitted or Overfitted Model 100

    Error Analysis 100

    Exercise 3.02: Performing Error Analysis 103

    Activity 3.02: Improving a Model's Performance 104

Deploying Your Model 106

    Saving and Loading Your Model 106

    PyTorch for Production in C++ 108

    Building an API 108

    Exercise 3.03: Creating a Web API 111

    Activity 3.03: Making Use of Your Model 113

Summary 116

Chapter 4: Convolutional Neural Networks 119

Introduction 120

Building a CNN 120

Why Are CNNs Used for Image Processing? 121

The Image as Input 122

Applications of CNNs 123

Classification 124

Localization 124

Detection 125

Segmentation 126

The Building Blocks of CNNs 126

Convolutional Layers...127

Exercise 4.01: Calculating the Output Shape
of a Convolutional Layer 132

Pooling Layers..134

Exercise 4.02: Calculating the Output Shape of a Set
of Convolutional and Pooling Layers 135

Fully Connected Layers.......................................137

Side Note – Downloading Datasets from PyTorch 140

Activity 4.01: Building a CNN for an Image
Classification Problem 142

Data Augmentation 144

Data Augmentation with PyTorch 145

Activity 4.02: Implementing Data Augmentation 147

Batch Normalization 149

Batch Normalization with PyTorch 151

Activity 4.03: Implementing Batch Normalization 152

Summary 154

Chapter 5: Style Transfer   157

Introduction 158

Style Transfer 159

   How Does It Work? 160

Implementation of Style Transfer Using the
VGG-19 Network Architecture 161

   Inputs – Loading and Displaying 162

   Exercise 5.01: Loading and Displaying Images 164

   Loading the Model 168

   Exercise 5.02: Loading a Pre-Trained Model in PyTorch 169

   Extracting the Features 170

   Exercise 5.03: Setting Up the Feature Extraction Process 173

   The Optimization Algorithm, Losses, and Parameter Update 176

      Content Loss..176

      Style Loss 176

      Total Loss...177

   Exercise 5.04: Creating the Target Image 178

   Activity 5.01: Performing Style Transfer 182

Summary 184

Chapter 6: Analyzing the Sequence of Data with RNNs   187

Introduction 188

Recurrent Neural Networks 188

   Applications of RNNs 189

How Do RNNs Work? 194

Input and Targets for Sequenced Data 197

Exercise 6.01: Creating the Input and Target Variables
for a Sequenced Data Problem 198

RNNs in PyTorch 202

Activity 6.01: Using a Simple RNN for a Time
Series Prediction 204

Long Short-Term Memory Networks 206

Applications of LSTM Networks 207

How Do LSTM Networks Work? 208

LSTM Networks in PyTorch 211

Preprocessing the Input Data 212

Numbered Labels 212

Generating the Batches 213

One-Hot Encoding 215

Exercise 6.02: Preprocessing the Input Data and Creating
a One-Hot Matrix 216

Building the Architecture 219

Training the Model 220

Performing Predictions 222

Activity 6.02: Text Generation with LSTM Networks 223

Natural Language Processing 225

Sentiment Analysis 226

Sentiment Analysis in PyTorch 227

Preprocessing the Input Data 228

Building the Architecture 229

Training the Model 229

Activity 6.03: Performing NLP for Sentiment Analysis 230

Summary 233

Appendix   235

Index   305

PREFACE

ABOUT THE BOOK

Want to get to grips with one of the most popular machine learning libraries for deep learning? *The Deep Learning with PyTorch Workshop* will help you do just that, jumpstarting your knowledge of using PyTorch for deep learning even if you're starting from scratch.

It's no surprise that deep learning's popularity has risen steeply in the past few years, thanks to intelligent applications such as self-driving vehicles, chatbots, and voice-activated assistants that are making our lives easier. This book will take you inside the world of deep learning, where you'll use PyTorch to understand the complexity of neural network architectures.

The Deep Learning with PyTorch Workshop starts with an introduction to deep learning and its applications. You'll explore the syntax of PyTorch and learn how to define a network architecture and train a model. Next, you'll learn about three main neural network architectures - convolutional, artificial, and recurrent - and even solve real-world data problems using these networks. Later chapters will show you how to create a style transfer model to develop a new image from two images, before finally taking you through how RNNs store memory to solve key data issues.

By the end of this book, you'll have mastered the essential concepts, tools, and libraries of PyTorch to develop your own deep neural networks and intelligent apps.

AUDIENCE

This deep learning book is ideal for anyone who wants to create and train deep learning models using PyTorch. A solid understanding of the Python programming language and its packages will help you grasp the topics covered in the book more quickly.

ABOUT THE CHAPTERS

Chapter 1, Introduction to Deep Learning and PyTorch, introduces deep learning and its applications as well as the main syntax of PyTorch. The chapter also shows how you can define a network architecture and train a model.

Chapter 2, Building Blocks of Neural Networks, introduces the concept of neural networks and explains the three main network architectures nowadays: artificial neural networks, convolutional neural networks, and recurrent neural networks. For each architecture, an explanation of the training process and the layers is provided.

Chapter 3, A Classification Problem Using DNNs, introduces a real-life data problem to be solved using an artificial neural network. The preprocessing of the dataset is explored, as well as the process of defining and training the model, and improving its accuracy through the use of error analysis.

Chapter 4, Convolutional Neural Networks, looks at convolutional neural networks in greater detail. Using a real-life data problem, you'll learn how to construct the network architecture and train it, as well as how to improve the results by using data augmentation and batch normalization.

Chapter 5, Style Transfer, demonstrates the process of performing the task of style transfer, where two images are taken as input to create a new image, with elements from both input images.

Chapter 6, Analyzing the Sequences of Data with RNNs, explores recurrent neural networks in greater detail. In the chapter, three popular data problems are solved using sequenced data as input.

CONVENTIONS

Code words in text, database table names, folder names, filenames, file extensions, pathnames, dummy URLs, user input, and Twitter handles are shown as follows:

"Import **torch**, the **optim** package from PyTorch, and **matplotlib**:"

Words that you see on the screen, for example, in menus or dialog boxes, also appear in the same format.

A block of code is set as follows:

```
import torch
import torch.optim as optim
import matplotlib.pyplot as plt
```

New terms and important words are shown like this: "The chapter will also explore the concept of **Natural Language Processing** (**NLP**)."

CODE PRESENTATION

Lines of code that span multiple lines are split using a backslash (\ ). When the code is executed, Python will ignore the backslash, and treat the code on the next line as a direct continuation of the current line.

For example:

```
history = model.fit(X, y, epochs=100, batch_size=5, verbose=1, \
                   validation_split=0.2, shuffle=False)
```

Comments are added into code to help explain specific bits of logic. Single-line comments are denoted using the # symbol, as follows:

```
# Print the sizes of the dataset
print("Number of Examples in the Dataset = ", X.shape[0])
print("Number of Features for each example = ", X.shape[1])
```

Multi-line comments are enclosed by triple quotes, as shown below:

```
"""
Define a seed for the random number generator to ensure the result will be
reproducible
"""
seed = 1
np.random.seed(seed)
random.set_seed(seed)
```

HARDWARE REQUIREMENTS

For the optimal student experience, we recommend the following hardware configuration:

- Processor: Intel Core i3 or equivalent
- Memory: 4 GB RAM
- Storage: 35 GB available space

SOFTWARE REQUIREMENTS

You'll also need the following software installed in advance:

- OS: Windows 7 SP1 64-bit, Windows 8.1 64-bit, or Windows 10 64-bit, Ubuntu Linux, or the latest version of macOS

- Browser: Google Chrome/Mozilla Firefox (the latest version)

- Notepad++/Sublime Text as an IDE (optional, as you can practice everything using Jupyter notebooks in your browser)

- Python 3.7 with libraries as required (Jupyter, NumPy, pandas, Matplotlib, pillow, flask, xlrd, and scikit-learn)

- PyTorch 1.3+ (preferably PyTorch 1.4, with or without CUDA)

SETTING UP YOUR ENVIRONMENT

Before we explore the book in detail, we need to set up specific software and tools. In the following section, we shall see how to do that.

INSTALLING PYTHON ON WINDOWS AND MACOS

1. Visit the following link to download Python 3.7: https://www.python.org/downloads/release/python-376/.

2. At the bottom of the page, locate the table under the heading **Files**:

   For Windows, click on **Windows x86-64 executable installer** for 64-bit or **Windows x86 executable installer** for 32-bit.

   For macOS, click on **macOS 64-bit/32-bit installer** for macOS X 10.6 and later, or **macOS 64-bit installer** for OS X 10.9 and later.

3. Run the installer that you have downloaded.

4. You can also install Python using the Anaconda distribution. Follow the instructions given in the this link for more details: https://www.anaconda.com/products/individual

INSTALLING PYTHON ON LINUX

1. Open your Terminal and type the following command:

```
sudo apt-get install python3.7
```

You can also install Python using the Anaconda distribution. Follow the instructions given in the this link for more details: https://www.anaconda.com/products/individual

INSTALLING PIP

pip is included by default with the installation of Python 3.7. However, it may be the case that it did not get installed. To check whether it was installed, execute the following command in your terminal or command prompt:

```
pip --version
```

You might need to use the **pip3** command, due to previous versions of **pip** on your computer that are already using the **pip** command.

If the **pip** (or **pip3**) command is not recognized by your machine, follow these steps to install it:

1. To install **pip**, visit the following link and download the **get-pip.py** file: https://pip.pypa.io/en/stable/installing/.

2. Then, on the Terminal or Command Prompt, use the following command to install it:

```
python get-pip.py
```

You might need to use the **python3 get-pip.py** command, due to previous versions of Python on your machine that are already using the **python** command.

INSTALLING PYTORCH

To install PyTorch, with or without CUDA, follow these steps:

1. Visit the following link: https://pytorch.org/get-started/locally/.

2. Under the **Start Locally** heading, select the options that apply to you. This will give you the command that you need to execute to download PyTorch on your local machine. Use **pip** as the package to download PyTorch.

3. Copy the command and run it in your Terminal or Command Prompt.

INSTALLING LIBRARIES

pip comes pre-installed with Anaconda. Once Anaconda is installed on your machine, all the required libraries can be installed using **pip**, for example, **pip install numpy**. Alternatively, you can install all the required libraries using **pip install -r requirements.txt**. You can find the **requirements.txt** file at https://packt.live/3ih86lh.

The exercises and activities will be executed in Jupyter Notebooks. Jupyter is a Python library and can be installed in the same way as the other Python libraries – that is, with **pip install jupyter**, but fortunately, it comes pre-installed with Anaconda. To open a notebook, simply run the command **jupyter notebook** in the Terminal or Command Prompt.

OPENING A JUPYTER NOTEBOOK

1. Open a Terminal/Command Prompt.

2. In the Terminal/Command Prompt, go to the directory location where you have downloaded the book's GitHub repository.

3. Open a Jupyter Notebook by typing in the following command:

    ```
    jupyter notebook
    ```

    By executing the previous command, you will be able to use Jupyter Notebooks through the default browser of your machine.

ACCESSING THE CODE FILES

You can find the complete code files of this book at https://packt.live/38qLadV. You can also run many activities and exercises directly in your web browser by using the interactive lab environment at https://packt.live/3ieFDg1.

We've tried to support interactive versions of all activities and exercises, but we recommend a local installation as well for instances where this support isn't available.

If you have any issues or questions about installation, please email us at workshops@packt.com.

1

INTRODUCTION TO DEEP LEARNING AND PYTORCH

OVERVIEW

This chapter introduces the two main topics of this book: deep learning and PyTorch. Here, you will be able to explore some of the most popular applications of deep learning, understand what PyTorch is, and use PyTorch to build a single-layer network, which will be the starting point for you to apply your learning to real-life data problems. By the end of this chapter, you will be able to use PyTorch's syntax to build neural networks, which will be essential in subsequent chapters.

INTRODUCTION

Deep learning is a subset of machine learning that focuses on using neural networks to solve complex data problems. It is becoming increasingly popular nowadays, thanks to advances in software and hardware that allow us to gather and process large amounts of data (we are talking about millions and billions of entries). This is important considering that deep neural networks require vast amounts of data to perform well.

Some of the most well-known applications of deep learning are self-driving vehicles, popular chatbots, and a wide variety of voice-activated assistants, which will be further explained in this chapter.

PyTorch was launched back in 2017, and its main characteristic is that it uses **graphics processing units** (**GPUs**) to process data using "tensors". This allows algorithms to run at high speeds and, at the same time, it provides its users with flexibility and a standard syntax to obtain the best results for many data problems. Furthermore, PyTorch uses dynamic computational graphs that allow you to make changes to the network on the go. This book demystifies neural networks using PyTorch and helps you understand the complexity of neural network architectures.

WHY DEEP LEARNING?

In this section, we will cover the importance of deep learning and the reasons for its popularity.

Deep learning is a subset of machine learning that uses multi-layer neural networks (large neural networks), inspired by the biological structure of the human brain, where neurons in a layer receive some input data, process it, and send the output to the following layer. These neural networks can consist of thousands of interconnected nodes (neurons), mostly organized in different layers, where one node is connected to several nodes in the previous layer from where it receives its input data, as well as being connected to several nodes in the following layer, to which it sends the output data after it has been processed.

Deep learning's popularity is due to its accuracy. It has achieved higher accuracy levels than other algorithms have ever before for complex data problems such as **natural language processing** (**NLP**). Deep learning's ability to perform outstandingly well has reached levels where machines can outperform humans, such as in the case of fraud detection. Deep learning models can not only optimize processes but also improve their quality. This has meant advances in revolutionary fields where accuracy is vital for safety reasons, such as self-driven cars.

Even though neural networks were theorized decades ago, there are two main reasons why they have recently become popular:

- Neural networks require, and actually capitalize on, vast amounts of labeled data to achieve an optimal solution. This means that for the algorithm to create an outstanding model, it requires hundreds of thousands or even millions of entries, containing both the features and the target values. For instance, as regards the image recognition of cats, the more images you have, the more features the model is able to detect, which makes it better.

> **NOTE**
>
> Labeled data refers to data that contains a set of features (characteristics that describe an instance) and a target value (the value to be achieved); for example, a dataset containing demographical and financial information, with a target feature that determines the wage of a person.

The following plot shows the performance of deep learning against other algorithms in terms of the quantity of data:

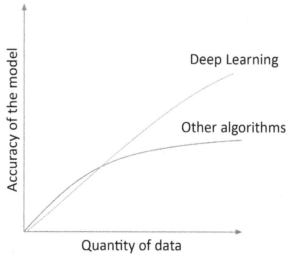

Figure 1.1: Performance of deep learning against other algorithms

This is possible nowadays thanks to advances in software and hardware that allow us to gather and process such granularity.

- Neural networks require considerable computing power to be able to process such large amounts of data without taking weeks (or even longer) to be trained. Because the process of achieving the best possible model is based on trial and error, it is necessary to be able to run the training process as efficiently as possible.

  This can be achieved today through the use of GPUs, which can cut down the training time of a neural network from weeks to hours.

> **NOTE**
>
> With the objective of accelerating deep learning in order to be able to make use of large amounts of training data and construct state-of-the-art models, **field-programmable gate arrays** (**FPGAs**) and **tensor processing units** (**TPUs**) are being developed by major cloud computing providers, such as AWS, Microsoft Azure, and Google.

APPLICATIONS OF DEEP LEARNING

Deep learning is revolutionizing technology and is already impacting our lives. Deep learning can be applied to a wide variety of situations, ranging from medical and safety (such as fraud detection) purposes to more trivial tasks, such as colorizing black and white images or translating text in real time.

Some of the applications of deep learning that are either under development or in use today include the following:

- **Self-driving vehicles**: Several companies, such as Google, have been working on the development of partially or totally self-driving vehicles that learn to drive by using digital sensors to identify the objects around them.

- **Medical diagnosis**: Deep learning is impacting this industry by improving the diagnosis accuracy of terminal diseases such as brain and breast cancer. This is done by classifying X-rays (or any other diagnostic imagery mechanisms) of new patients, based on labeled X-rays from previous patients that did or did not have cancer.

- **Voice assistants**: This may be one of the most popular applications nowadays, due to the proliferation of different voice-activated intelligent assistants, such as Apple's Siri, Google Home, and Amazon's Alexa.

- **Automatic text generation**: This means generating new text based on an input sentence. This is popularly used in email writing, where the email provider suggests the next couple of words to the user, based on the text that's already been written.

- **Advertising**: In the commercial world, deep learning is helping to increase the return on investment of advertising campaigns by targeting the right audiences and by creating more effective ads. One example of this is the generation of content in order to produce up-to-date and informative blogs that help to engage current customers and attract new ones.

- **Price forecasting**: For beginners, this is a typical example of what can be achieved through the use of machine learning algorithms. Price forecasting consists of training a model based on real data. For instance, in the field of real estate, this would consist of feeding a model with property characteristics and their final price in order to be able to predict the prices of future entries based solely on property characteristics.

INTRODUCTION TO PYTORCH

PyTorch is an open source library developed mainly by Facebook's artificial intelligence research group as a Python version of Torch.

> **NOTE**
>
> Torch is an open source, scientific computing framework that supports a wide variety of machine learning algorithms.

PyTorch was first released to the public in January 2017. It uses the power of **GPUs** to speed up the computation of tensors, which accelerates the training times of complex models.

The library has a C++ backend, combined with the deep learning framework of Torch, which allows much faster computations than native Python libraries with many deep learning features. The frontend is in Python, which has helped it gain popularity, enabling data scientists new to the library to construct complex neural networks. It is possible to use PyTorch alongside other popular Python packages.

Although the PyTorch is fairly new, it has gained popularity quickly as it was developed using feedback from many experts in the field. This has led PyTorch to become a useful library for users.

GPUS IN PYTORCH

GPUs were originally developed to speed up computations in graphics rendering, especially for video games and such. However, they have become increasingly popular lately thanks to their ability to help speed up computations for any field, including deep learning calculations.

There are several platforms that allow the allocation of variables to the GPUs of a machine, with the **Compute Unified Device Architecture** (**CUDA**) being one of the most commonly used platforms. CUDA is a computing platform developed by Nvidia that speeds up compute-intensive programs thanks to the use of GPUs to perform computations.

In PyTorch, the allocation of variables to CUDA can be done through the use of the `torch.cuda` package, as shown in the following code snippet:

```
x = torch.Tensor(10).random_(0, 10)
x.to("cuda")
```

Here, the first line of code creates a tensor filled with random integers (between 0 and 10). The second line of code allocates that tensor to CUDA so that all computations involving that tensor are handled by the GPU instead of the CPU. To allocate a variable back to the CPU, use the following code snippet:

```
x.to("cpu")
```

In CUDA, when solving a deep learning data problem, it is good practice to allocate the model holding the network architecture, as well as the input data. This will ensure that all computations carried out during the training process are handled by the GPU.

Nevertheless, this allocation can only be done given that your machine has a GPU available and that you have installed PyTorch with the CUDA package. To verify whether you are able to allocate your variables in CUDA, use the following code snippet:

```
torch.cuda.is_available()
```

If the output from the preceding line of code is **True**, you are all set to start allocating your variables in CUDA.

> **NOTE**
>
> To install PyTorch along with the CUDA package, visit PyTorch's website and make sure you select an option that includes CUDA (either version): https://pytorch.org/get-started/locally/.

WHAT ARE TENSORS?

Similar to NumPy, PyTorch uses tensors to represent data. Tensors are matrix-like structures of n dimensions with the difference being that PyTorch tensors can run on the GPU (while NumPy tensors cannot), which helps to accelerate numerical computations. For tensors, dimensions are also known as ranks. The following diagram shows a visual representation of tensors of different dimensions:

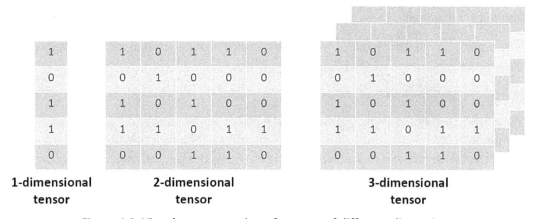

Figure 1.2: Visual representation of tensors of different dimensions

In contrast to a matrix, a tensor is a mathematical entity contained in a structure that can interact with other mathematical entities. When one tensor transforms another, the former also carries a transformation of its own.

This means that tensors are not just data structures, but rather containers that, when fed some data, can map in a multi-linear manner with other tensors.

Similar to NumPy arrays or any other matrix-like structure, PyTorch tensors can have as many dimensions as desired. Defining a one-dimensional tensor (**tensor_1**) and a two-dimensional tensor (**tensor_2**) in PyTorch can be achieved using the following code snippet:

```
tensor_1 = torch.tensor([1,1,0,2])
tensor_2 = torch.tensor([[0,0,2,1,2],[1,0,2,2,0]])
```

Note that the numbers in the preceding code snippet do not have a meaning. What matters is the definition of the different dimensions, which are filled with random numbers. From the preceding snippet, the first tensor would have a size of 4 for one dimension, while the second one would have a size of 5 for each of the two dimensions, which can be verified by making use of the **shape** property over the tensor variables, as seen here:

```
tensor_1.shape
```

The output is **torch.Size([4])**.

```
tensor_2.shape
```

The output is **torch.Size([2],[5])**.

When using a GPU-enabled machine, the following modification is implemented to define a tensor:

```
tensor = torch.tensor([1,1,0,2]).cuda()
```

Creating dummy data using PyTorch tensors is fairly simple, similar to what you would do in NumPy. For instance, **torch.randn()** returns a tensor filled with random numbers of the dimensions specified within the parentheses, while **torch.randint()** returns a tensor filled with integers (the minimum and maximum values can be defined) of the dimensions defined within the parentheses:

> **NOTE**
>
> The code snippet shown here uses a backslash (\ ) to split the logic across multiple lines. When the code is executed, Python will ignore the backslash, and treat the code on the next line as a direct continuation of the current line.

```
example_1 = torch.randn(3,3)
example_2 = torch.randint(low=0, high=2, \
                  size=(3,3)).type(torch.FloatTensor)
```

As can be seen, **example_1** is a two-dimensional tensor filled with random numbers, with each dimension of size equal to 3, while **example_2** is a two-dimensional tensor filled with 0s and 1s (the **high** parameter is upper-bound exclusive), with each dimension's size equal to 3.

Any tensor filled with integers must be converted into floats so that we can feed it to any PyTorch model.

EXERCISE 1.01: CREATING TENSORS OF DIFFERENT RANKS USING PYTORCH

In this exercise, we will use the PyTorch library to create tensors of ranks one, two, and three. Perform the following steps to complete this exercise:

> **NOTE**
>
> For the exercises and activities in this chapter, you will need to have Python 3.7, Jupyter 6.0, Matplotlib 3.1, and PyTorch 1.3+ (preferably PyTorch 1.4, with or without CUDA) installed (as instructed in the *Preface*). They will be primarily developed in a Jupyter Notebook and it is recommended that you keep a separate notebook for different assignments unless advised not to.

1. Import the PyTorch library called **torch**:

```
import torch
```

2. Create tensors of the following ranks: **1**, **2**, and **3**.

    Use values between **0** and **1** to fill your tensors. The size of the tensors can be defined as you wish, given that the ranks are created correctly:

```
tensor_1 = torch.tensor([0.1,1,0.9,0.7,0.3])
tensor_2 = torch.tensor([[0,0.2,0.4,0.6],[1,0.8,0.6,0.4]])
tensor_3 = torch.tensor([[[0.3,0.6],[1,0]], \
                         [[0.3,0.6],[0,1]]])
```

If your machine has a GPU available, you can create equivalent tensors using the GPU syntax:

```
tensor_1 = torch.tensor([0.1,1,0.9,0.7,0.3]).cuda()
tensor_2 = torch.tensor([[0,0.2,0.4,0.6], \
                        [1,0.8,0.6,0.4]]).cuda()
tensor_3 = torch.tensor([[[0.3,0.6],[1,0]], \
                         [[0.3,0.6],[0,1]]]).cuda()
```

3. Print the shape of each of the tensors using the **shape** property, just as you would do with NumPy arrays:

```
print(tensor_1.shape)
print(tensor_2.shape)
print(tensor_3.shape)
```

The output of the **print** statements should look as follows, considering that the size of each dimension of the tensors may vary according to your choices:

```
torch.Size([5])
torch.Size([2, 4])
torch.Size([2, 2, 2])
```

> **NOTE**
>
> To access the source code for this specific section, please refer to https://packt.live/3dOS66H.
>
> You can also run this example online at https://packt.live/2VwTLHq. You must execute the entire Notebook in order to get the desired result.
>
> To access the GPU version of this source code, please refer to https://packt.live/31AwIzo. This version of the source code is not available as an online interactive example, and will need to be run locally with the GPU setup.

You have successfully created tensors of different ranks.

In the next section, we will discuss the advantages and disadvantages of using PyTorch.

ADVANTAGES OF USING PYTORCH

There are several libraries nowadays that can be used to develop deep learning solutions, so why use PyTorch? The answer is that PyTorch is a dynamic library that allows its users great flexibility to develop complex architectures that can be adapted to a particular data problem.

PyTorch has been adopted by many researchers and artificial intelligence developers, which makes it an important tool to have in a machine learning engineers toolkit.

The key aspects to highlight are as follows:

- **Ease of use**: With respect to the API, PyTorch has a simple interface that makes it easy to develop and run models. Many early adopters consider it to be more intuitive than other libraries, such as TensorFlow.

- **Speed**: The use of GPUs enables the library to train faster than other deep learning libraries. This is especially useful when different approximations have to be tested in order to achieve the best possible model. Additionally, even though other libraries may also have the option to accelerate computations with GPUs, you can do this in PyTorch by typing just a couple of simple lines of code.

- **Convenience**: PyTorch is flexible. It uses dynamic computational graphs that allow you to make changes to networks on the go. It also allows great flexibility when building the architecture as it is easy to make adjustments to conventional architectures.

- **Imperative**: PyTorch is also imperative. Each line of code is executed individually, allowing you to track the model in real time, as well as debug the model in a convenient way.

- **Pretrained models**: Finally, it contains many pretrained models that are easy to use and are a great starting point for some data problems.

DISADVANTAGES OF USING PYTORCH

Although the advantages are huge and many, there are still some disadvantages to consider, which are explained here:

- **Small community**: The community of adapters of this library is small in comparison to other libraries, such as TensorFlow. However, having been available to the public for only 3 years, today, it is among the list of the top five most popular libraries for implementing deep learning solutions, and its community is growing by the day.

- **Spotty documentation**: Considering that the library is fairly new in comparison to other deep learning libraries, the documentation is not as complete. However, since the features and capabilities of the library are increasing, the documentation is being extended. Additionally, as the community continues to grow, there will be more information available on the internet.

- **Questions around production-readiness**: Although many of the complaints about the library have focused on its inability to be deployed for production, after the launch of version 1.0, the library has included production capabilities to be able to export finalized models and use them in production environments.

KEY ELEMENTS OF PYTORCH

Like any other library, PyTorch has a variety of modules, libraries, and packages for developing different functionalities. In this section, the three most commonly used elements for building deep neural networks will be explained, along with a simple example of the syntax.

THE PYTORCH AUTOGRAD LIBRARY

The **autograd** library consists of a technique called automatic differentiation. Its purpose is to numerically calculate the derivative of a function. This is crucial for a concept we will learn about in the next chapter called backward propagation, which is carried out while training a neural network.

The derivative (also known as the gradient) of an element refers to the rate of change of that element in a given time step. In deep learning, gradients refer to the dimension and magnitude in which the parameters of the neural network must be updated in a training step in order to minimize the loss function. This concept will be further explored in the following chapter.

> **NOTE**
>
> A detailed explanation of neural networks and the different steps taken to train a model will be given in subsequent sections.

To compute the gradients, simply call the **backward()** function, as shown here:

```
a = torch.tensor([5.0, 3.0], requires_grad=True)
b = torch.tensor([1.0, 4.0])
ab = ((a + b) ** 2).sum()
ab.backward()
```

In the preceding code, two tensors were created. We use the **requires_grad** argument here to tell PyTorch to calculate the gradients of that tensor. However, when building your neural network, this argument is not required.

Next, a function was defined using the values of both tensors. Finally, the **backward()** function was used to calculate the gradients.

By printing the gradients for both **a** and **b**, it is possible to confirm that they were only calculated for the first variable (**a**), while for the second one (**b**), it throws an error:

```
print(a.grad.data)
```

The output is **tensor([12., 14.])**.

```
print(b.grad.data)
```

The output is as follows:

```
AttributeError: 'NoneType' object has no attribute 'data'
```

THE PYTORCH NN MODULE

The **autograd** library alone can be used to build simple neural networks, considering that the trickier part (the calculation of gradients) has been taken care of. However, this methodology can be troublesome, hence the introduction of the **nn** module.

The **nn** module is a complete PyTorch module used to create and train neural networks, which, through the use of different elements, allows for simple and complex developments. For instance, the **Sequential()** container allows for the easy creation of network architectures that follow a sequence of predefined modules (or layers) without the need for much knowledge of defining network architectures.

> **NOTE**
> The different layers that can be used for each neural network architecture will be explained further in subsequent chapters.

This module also has the capability to define the loss function to evaluate the model and many more advanced features that will be discussed in this book.

The process of building a neural network architecture as a sequence of predefined modules can be achieved in just a couple of lines, as shown here:

```
import torch.nn as nn
model = nn.Sequential(nn.Linear(input_units, hidden_units), \
                      nn.ReLU(), \
                      nn.Linear(hidden_units, output_units), \
                      nn.Sigmoid())
loss_funct = nn.MSELoss()
```

First, the module is imported. And then, the model architecture is defined. **input_units** refers to the number of features that the input data contains, **hidden_units** refers to the number of nodes of the hidden layer, and **output_units** refers to the number of nodes of the output layer.

As can be seen in the preceding code, the architecture of the network contains one hidden layer, followed by a ReLU activation function and an output layer, followed by a sigmoid activation function, making it a two-layer network.

Finally, the loss function is defined as the **Mean Squared Error** (**MSE**).

> **NOTE**
>
> The most popular loss functions for different data problems will be explained throughout this book.
>
> To create models that do not follow a sequence of existing modules, **custom nn** modules are used. We'll introduce these later in this book.

EXERCISE 1.02: DEFINING A SINGLE-LAYER ARCHITECTURE

In this exercise, we will use PyTorch's **nn** module to define a model for a single-layer neural network, and also define the loss function to evaluate the model. This will be the starting point so that you will be able to build more complex network architectures to solve real-life data problems. Perform the following steps to complete this exercise:

1. Import **torch** as well as the **nn** module from PyTorch:

```
import torch
import torch.nn as nn
```

> **NOTE**
>
> **torch.manual_seed(0)** is being used in this exercise in order to ensure the reproducibility of the results that were obtained in this book's GitHub repository. However, when training a network for other purposes, a seed must not be defined.
>
> To learn more about seed in PyTorch,
> visit https://pytorch.org/docs/stable/notes/randomness.html.

2. Define the number of features of the input data as **10** (**input_units**) and the number of nodes of the output layer as **1** (**output_units**):

```
input_units = 10
output_units = 1
```

3. Using the **Sequential()** container, define a single-layer network architecture and store it in a variable named **model**. Make sure to define one layer, followed by a **Sigmoid** activation function:

```
model = nn.Sequential(nn.Linear(input_units, output_units), \
                      nn.Sigmoid())
```

4. Print your model to verify that it was created accordingly:

```
print(model)
```

Running the preceding code snippet will display the following output:

```
Sequential(
    (0): Linear(in_features=10, out_features=1, bias=True)
    (1): Sigmoid()
)
```

5. Define the loss function as the MSE and store it in a variable named **loss_funct**:

```
loss_funct = nn.MSELoss()
```

6. Print your loss function to verify that it was created accordingly:

```
print(loss_funct)
```

Running the preceding code snippet will display the following output:

```
MSELoss()
```

> **NOTE**
>
> To access the source code for this specific section, please refer to https://packt.live/2YNwyTy.
>
> You can also run this example online at https://packt.live/2YOVPws. You must execute the entire Notebook in order to get the desired result.

You have successfully defined a single-layer network architecture.

THE PYTORCH OPTIM PACKAGE

The **optim** package is used to define the optimizer that will be used to update the parameters in each iteration (which will be further explained in the following chapters) using the gradients calculated by the **autograd** module. Here, it is possible to choose from different optimization algorithms that are available, such as **Adam**, **Stochastic Gradient Descent (SGD)**, and **Root Mean Square Propagation (RMSprop)**, among others.

> **NOTE**
>
> The most popular optimization algorithms will be explained in subsequent chapters.

To set the optimizer to be used, the following line of code shall suffice, after importing the package:

```
optimizer = torch.optim.SGD(model.parameters(), lr=0.01)
```

Here, the **model.parameters()** argument refers to the weights and biases from the model that were previously created, while **lr** refers to the learning rate, which was set to **0.01**.

Weights are the values that are used to determine the level of importance of a bit of information in a general context. This means that every bit of information has an accompanying weight for every neuron in the network. Moreover, bias is similar to the intercept element that's added to a linear function and is used to adjust the output from the computation of relevance in a given neuron.

The learning rate is a running parameter that's used in optimization processes to determine the extent of the steps to be taken toward minimizing the loss function.

Next, the process of running the optimization for 100 iterations is shown here, which, as you can see, uses the model created by the **nn** module and the gradients calculated by the **autograd** library:

> **NOTE**
>
> The # symbol in the code snippet below denotes a code comment. Comments are added into code to help explain specific bits of logic. The triple-quotes (**"""**) shown in the code snippet below are used to denote the start and end points of a multi-line code comment. Comments are added into code to help explain specific bits of logic.

```
for i in range(100):

    # Call to the model to perform a prediction
    y_pred = model(x)

    # Calculation of loss function based on y_pred and y
    loss = loss_funct(y_pred, y)

    # Zero the gradients so that previous ones don't accumulate
    optimizer.zero_grad()

    # Calculate the gradients of the loss function
    loss.backward()

    """
    Call to the optimizer to perform an update
    of the parameters
    """
    optimizer.step()
```

For each iteration, the model is called to obtain a prediction (**y_pred**). This prediction and the ground truth values (**y**) are fed to the loss functions in order to determine the ability of the model to approximate to the ground truth.

Next, the gradients are zeroed, and the gradients of the loss function are calculated using the **backward()** function.

Finally, the **step()** function is called to update the weights and biases based on the optimization algorithm and the gradients calculated previously.

EXERCISE 1.03: TRAINING A NEURAL NETWORK

> **NOTE**
>
> For this exercise, use the same Jupyter Notebook from the previous exercise (*Exercise 1.02*, *Defining a Single-Layer Architecture*).

In this exercise, we will learn how to train the single-layer network from the previous exercise, using PyTorch's **optim** package. Considering that we will use dummy data as input, training the network won't solve a data problem, but it will be performed for learning purposes. Perform the following steps to complete this exercise:

1. Import **torch**, the **optim** package from PyTorch, and **matplotlib**:

```
import torch
import torch.optim as optim
import matplotlib.pyplot as plt
```

2. Create dummy input data (**x**) of random values and dummy target data (**y**) that only contains zeros and ones. Tensor **x** should have a size of (**20 , 10**), while the size of **y** should be (**20 , 1**):

```
x = torch.randn(20,10)
y = torch.randint(0,2, (20,1)).type(torch.FloatTensor)
```

3. Define the optimization algorithm as the Adam optimizer. Set the learning rate equal to **0.01**:

```
optimizer = optim.Adam(model.parameters(), lr=0.01)
```

4. Run the optimization for 20 iterations, saving the value of the loss in a variable. Every five iterations, print the loss value:

```
losses = []
for i in range(20):
    y_pred = model(x)
    loss = loss_funct(y_pred, y)
    losses.append(loss.item())
    optimizer.zero_grad()
    loss.backward()
```

```
        optimizer.step()

        if i%5 == 0:
            print(i, loss.item())
```

The output should look as follows:

```
0  0.25244325399398804
5  0.23448510468006134
10  0.21932794153690338
15  0.20741790533065796
```

The preceding output displays the epoch number, as well as the value for the loss function, which, as can be seen, is decreasing. This means that the training process is minimizing the loss function, which means that the model is able to understand the relationship between the input features and the target.

5. Make a line plot to display the value of the loss function in each epoch:

```
plt.plot(range(0,20), losses)
plt.show()
```

The output should look as follows:

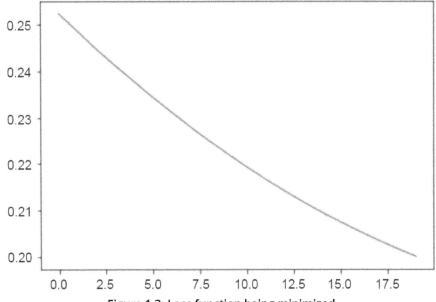

Figure 1.3: Loss function being minimized

As you can see, the loss function is being minimized.

> **NOTE**
>
> To access the source code for this specific section, please refer to https://packt.live/2NJrPfd.
>
> You can also run this example online at https://packt.live/2BTnXWw. You must execute the entire Notebook in order to get the desired result.

With that, you have successfully trained a single-layer neural network.

ACTIVITY 1.01: CREATING A SINGLE-LAYER NEURAL NETWORK

For this activity, we will create a single-layer neural network, which will be a starting point from which we will create deep neural networks in future activities. Let's look at the following scenario.

You work as an assistant of the mayor of Somerville and the HR department has asked you to build a model capable of predicting whether a person is happy with the current administration based on their satisfaction with the city's services. To do so, you have decided to build a single-layer neural network using PyTorch, using the response of previous surveys. Perform the following steps to complete this activity:

> **NOTE**
>
> The dataset that's being used for this activity was taken from the UC Irvine Machine Learning Repository, which can be downloaded using the following URL, from the **Data Folder** hyperlink: https://archive.ics.uci.edu/ml/datasets/Somerville+Happiness+Survey. It is also available in this book's GitHub repository: https://packt.live/38gzpr5.

1. Import the required libraries, including pandas for reading a CSV file.

2. Read the CSV file containing the dataset.

> **NOTE**
>
> It is recommended to use pandas' **read_csv** function to load the CSV file. To find out more about this function, visit https://pandas.pydata.org/pandas-docs/stable/reference/api/pandas.read_csv.html.

3. Separate the input features from the target. Note that the target is located in the first column of the CSV file. Next, convert the values into tensors, making sure the values are converted into floats.

> **NOTE**
>
> To slice a pandas DataFrame, use pandas' **iloc** method. To find out more about this method, visit https://pandas.pydata.org/pandas-docs/stable/reference/api/pandas.DataFrame.iloc.html.

4. Define the architecture of the model and store it in a variable named **model**. Remember to create a single-layer model.

5. Define the loss function to be used. In this case, use the MSE loss function.

6. Define the optimizer of your model. In this case, use the Adam optimizer and a learning rate of **0.01**.

7. Run the optimization for 100 iterations, saving the loss value for each iteration. Print the loss value every 10 iterations.

8. Make a line plot to display the loss value for each iteration step.

> **NOTE**
>
> The solution to this activity can be found on page 236.

SUMMARY

Deep learning is a subset of machine learning that was inspired by the biological structure of human brains. It uses deep neural networks to solve complex data problems through the use of vast amounts of data. Even though the theory was developed decades ago, it has been used recently thanks to advances in hardware and software that allow us to collect and process millions of pieces of data.

With the popularity of deep learning solutions, many deep learning libraries have been developed. Among them, one of the most recent ones is PyTorch. PyTorch uses a C++ backend, which helps speed up computation, while having a Python frontend to keep the library easy to use.

It uses tensors to store data, which are n-ranked matrix-like structures that can be run on GPUs to speed up processing. It offers three main elements that are highly useful for creating complex neural network architectures with little effort.

The **autograd** library can compute the derivatives of a function, which are used as the gradients to optimize the weights and biases of a model. Moreover, the **nn** module helps you to easily define the model's architecture as a sequence of predefined modules, as well as to determine the loss function to be used to measure the model. Finally, the **optim** package is used to select the optimization algorithm to be used to update the parameters, considering the gradients calculated previously.

In the next chapter, we will learn about the building blocks of a neural network. We will cover the three types of learning processes, as well as the three most common types of neural networks. For each neural network, we will learn how the network architecture is structured, as well as how the training process works. Finally, we will learn about the importance of data preparation and solve a regression data problem.

2

BUILDING BLOCKS OF NEURAL NETWORKS

OVERVIEW

This chapter introduces the main building blocks of neural networks and also explains the three main neural network architectures nowadays. Moreover, it explains the importance of data preparation before training any artificial intelligence model, and finally explains the process of solving a regression data problem. By the end of this chapter, you will have a firm grasp of the learning process of different network architectures and their different applications.

INTRODUCTION

In the previous chapter, it was explained why deep learning has become so popular nowadays, and PyTorch was introduced as one of the most popular libraries for developing deep learning solutions. Although the main syntax for building a neural network using PyTorch was explained, in this chapter, we will further explore the concept of neural networks.

Although neural network theory was developed several decades ago, since the concept evolved from the notion of the perceptron, different architectures have been created to solve different data problems in recent times. This is, in part, due to the different data formats that can be found in real-life data problems, such as text, audio, and images.

The purpose of this chapter is to dive into the topic of neural networks and their main advantages and disadvantages so that you can understand when and how to use them. Then, we will explain the building blocks of the most popular neural network architectures: **artificial neural networks** (**ANNs**), **convolutional neural networks** (**CNNs**), and **recurrent neural networks** (**RNNs**).

Following this, the process of building an effective model will be explained by solving a real-life regression problem. This includes preparing the data to be fed to the neural network (also known as data preprocessing), defining the neural network architecture to be used, and evaluating the performance of the model, with the objective of determining how it can be improved to achieve an optimal solution.

The aforementioned process will be done using one of the neural network architectures that will be discussed in this chapter, all while taking into consideration that the solution for each data problem should be carried out using the architecture that performs best for the data type in question. The other architectures will be used in subsequent chapters to solve more complicated data problems that involve using images and sequences of text as input data.

> **NOTE**
>
> All the code present in this chapter can be found at:
> https://packt.live/34MBauE.

INTRODUCTION TO NEURAL NETWORKS

Neural networks learn from training data, rather than being programmed to solve a particular task by following a set of rules. This learning process can follow one of the following methodologies:

- **Supervised learning**: This is the simplest form of learning as it consists of a labeled dataset, where the neural network finds patterns that explain the relationship between the features and the target. The iterations during the learning process aim to minimize the difference between the predicted value and the ground truth. One example of this is classifying a plant based on the attributes of its leaves.

- **Unsupervised learning**: In contrast to the preceding methodology, unsupervised learning consists of training a model with unlabeled data (meaning that there is no target value). The purpose of this is to arrive at a better understanding of the input data. In general, networks take input data, encode it, and then reconstruct the content from the encoded version, ideally keeping the relevant information. For instance, given a paragraph, a neural network can map the words and then suggest which ones are the most important or descriptive for the paragraph. These can then be used as tags.

- **Reinforcement learning**: This methodology consists of learning from the input data, with the main objective of maximizing a reward function in the long run. This is achieved by learning from the data as it comes in, rather than it being trained over static data (as in supervised learning). Hence, decisions are not made based on the immediate reward, but on the accumulation of it in the entire learning process. An example of this is a model that allocates resources to different tasks, with the objective of minimizing bottlenecks that slow down general performance.

> **NOTE**
>
> From the learning methodologies we've mentioned here, the most commonly used one is supervised learning, which is the one that will be mainly used in subsequent sections. This means that all the exercises, activities, and examples in this chapter will use a labeled dataset as input data.

WHAT ARE NEURAL NETWORKS?

As we discussed earlier, neural networks are a type of machine learning algorithm that's modeled on the anatomy of the human brain and that uses mathematical equations to learn a pattern from the observations that were made from the training data.

However, to actually understand the logic behind the training process that neural networks typically follow, it is important to understand the concept of perceptrons.

Developed during the 1950s by Frank Rosenblatt, a perceptron is an artificial neuron that takes several inputs and produces a binary output, similar to neurons in the human brain. This then becomes the input of a subsequent perceptron (neuron). Perceptrons are the essential building blocks of a neural network (just like neurons are the building blocks of the human brain):

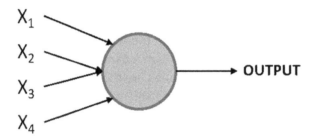

Figure 2.1: Diagram of a perceptron

Here, X_1, X_2, X_3, and X_4 represent the different inputs of the perceptron, and there could be any number of these. The circle is the perceptron, which is where the inputs are processed to arrive at an output.

Rosenblatt also introduced the concept of weights (w_1, w_2, ..., w_n), which are numbers that express the importance of each input. The output can be either 0 or 1, and it depends on whether the weighted sum of the inputs is above or below a given threshold (a numerical limit set by the developer or by a constraint of the data problem), which can be set as a parameter of the perceptron, as can be seen here:

$$output = \begin{cases} 0 \ if \ \sum_i w_i x_i \ > \ threshold \\ 1 \ if \ \sum_i w_i x_i \ \leq \ threshold \end{cases}$$

Figure 2.2: Equation for the output of perceptrons

EXERCISE 2.01: PERFORMING THE CALCULATIONS OF A PERCEPTRON

The following exercise does not require programming of any kind; instead, it consists of simple calculations to help you understand the notion of the perceptron. To perform these calculations, consider the following scenario.

There is a music festival in your town next Friday, but you are ill and trying to decide whether to go (where 0 means you are not going and 1 means you are going). Your decision relies on three factors:

- Will there be good weather? (X_1)

- Do you have someone to go with? (X_2)

- Is the music to your liking? (X_3)

For the preceding factors, we will use 1 if the answer to the question is yes, and 0 if the answer is no. Additionally, since you are very sick, the factor related to the weather is highly relevant, and you decide to give this factor a weight twice as big as the other two factors. Hence, you decide that the weights for the factors will be 4 (w_1), 2 (w_2), and 2 (w_3). Now, consider a threshold of 5:

1. With the information provided, calculate the output of the perceptron when considering that the weather is not good next Friday, but that you have someone to go with and you like the music at the festival:

$$ouput = X_1{}^*w_1 + X_2{}^*w_2 + X_3{}^*w_3$$

$$output = 0{}^*4 + 1{}^*2 + 1{}^*2 = 4$$

Figure 2.3: Output of the perceptron

Considering that the output is less than the threshold, the final result will be equal to 0, meaning that you should not go to the festival to avoid the risk of getting even more ill.

You have successfully performed the calculations of a perceptron, which is the starting point of understanding the learning process that occurs inside neural networks.

MULTI-LAYER PERCEPTRON

Considering what we learned about in the previous section, the notion of a multi-layered network consists of a network of multiple perceptrons stacked together (also known as nodes or neurons), such as the one shown here:

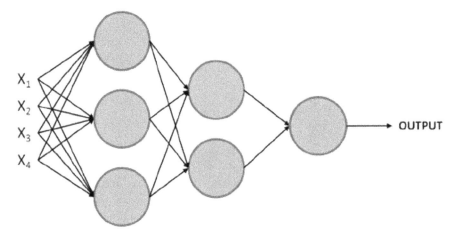

Figure 2.4: Diagram of a multi-layer perceptron

NOTE

The conventional way to refer to the layers in a neural network is as follows:

The first layer is the input layer, the last layer is the output layer, and all the layers in between are hidden layers.

Here, again, a set of inputs is used to train the model, but instead of feeding a single perceptron, they are fed to all the perceptrons (neurons) in the first layer. Next, the outputs that are obtained from this layer are used as inputs for the perceptrons in the subsequent layer and so on until the final layer is reached, which is in charge of outputting a result.

Note that the first layer of a perceptron handles a simple decision process by weighting the inputs, while the subsequent layer can handle more complex and abstract decisions based on the output of the previous layer, and hence the state-of-the-art performance of deep neural networks (networks that use many layers) for complex data problems.

Different to conventional perceptrons, neural networks have evolved to have one or multiple nodes in the output layer so that they are able to present the result either as binary or multiclass.

THE LEARNING PROCESS OF A NEURAL NETWORK

In general terms, a neural network is made up of multiple neurons, where each neuron computes a linear function, along with an activation function, to arrive at an output based on some inputs (an activation function is designed to break linearity – this will be explained in more detail later in this chapter). This output is tied to a weight, which represents its level of importance, and will be used for calculations in the following layer.

Moreover, these calculations are carried out throughout the entire architecture of the network, until a final output is reached. This output is used to determine the performance of the network in comparison to the ground truth, which is then used to adjust the different parameters of the network to start the calculation process over again.

Considering this, the training process of a neural network can be seen as an iterative process that goes forward and backward through the layers of the network to arrive at an optimal result, which can be seen in the following diagram (loss functions will be covered later in this chapter):

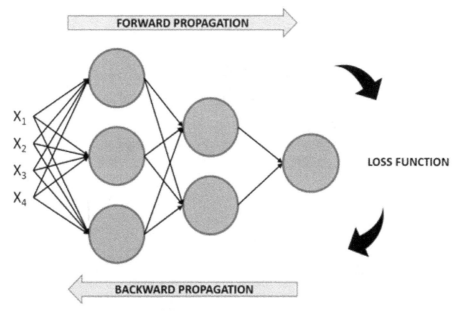

Figure 2.5: Diagram of the learning process of a neural network

FORWARD PROPAGATION

This is the process of going from left to right through the architecture of the network while performing calculations using the input data to arrive at a prediction that can be compared to the ground truth. This means that every neuron in the network will transform the input data (the initial data or data received from the previous layer) according to the weights and biases that it has associated with it and will send the output to the subsequent layer until a final layer is reached and a prediction is made.

> NOTE
>
> In neural networks, biases are numerical values that help shift the activation function of each neuron in order to avoid zero values that may affect the training process. Their role in the training of neural networks will be explained later in this chapter.

The calculations that are performed in each neuron include a linear function that multiplies the input data by some weight plus a bias, which is then passed through an activation function. The main purpose of the activation function is to break the linearity of the model, which is crucial considering that most real-life data problems that are solved using neural networks are not defined by a line, but rather by a complex function. These formulas are as follows:

$$Z = X*W + b$$
$$A = \sigma(Z)$$

Figure 2.6: Calculations performed by each neuron

Here, as we mentioned previously, X refers to the input data, W is the weight that determines the level of importance of the input data, b is the bias value, and sigma (σ) represents the activation function that's applied over the linear function.

The activation function serves the purpose of introducing non-linearity to the model. There are different activation functions to choose from, and a list of the ones most commonly used nowadays is as follows:

- **Sigmoid**: This is S-shaped, and it basically converts values into simple probabilities between 0 and 1, where most of the outputs that are obtained by the sigmoid function will be close to the extremes of 0 and 1:

$$\sigma(z) = \frac{1}{(1 + e^{-z})}$$

Figure 2.7: Sigmoid activation function

The following plot shows the graphical representation of the sigmoid activation function:

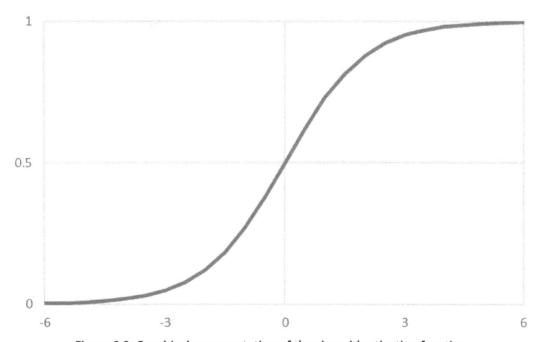

Figure 2.8: Graphical representation of the sigmoid activation function

- **Softmax**: Similar to the sigmoid function, this calculates the probability distribution of an event over *n* events, meaning that its output is not binary. In simple terms, this function calculates the probability of the output being one of the target classes in comparison to the other classes:

$$\sigma(Z) = \frac{e^z}{\sum\limits_{i=0}^{n} e^{z_i}}$$

Figure 2.9: Softmax activation function

Considering that its output is a probability, this activation function is often found in the output layer of classification networks.

- **Tanh**: This function represents the relationship between the hyperbolic sine and the hyperbolic cosine, and the result is between -1 and 1. The main advantage of this activation function is that negative values can be dealt with more easily:

$$\sigma = \frac{sinh(z)}{cosh(z)}$$

Figure 2.10: Tanh activation function

The following plot shows the graphical representation of the **tanh** activation function:

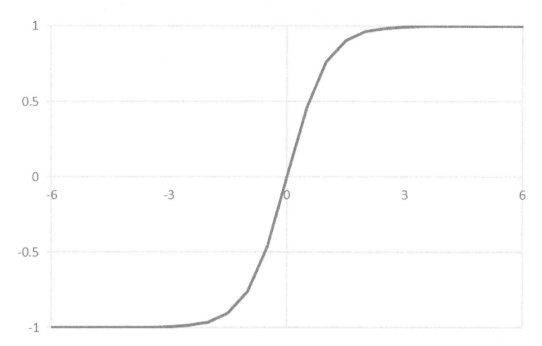

Figure 2.11: Graphical representation of the tanh activation function

- **Rectified Linear Function (ReLU):** This basically activates a node given that the output of the linear function is above 0; otherwise, its output will be 0. If the output of the linear function is above 0, the result from this activation function will be the raw number it received as input:

$$\sigma(z) = max(z, 0)$$

Figure 2.12: ReLU activation function

Conventionally, this activation function is used for all hidden layers. We will learn more about hidden layers in the upcoming sections of this chapter. The following plot shows the graphical representation of the ReLU activation function:

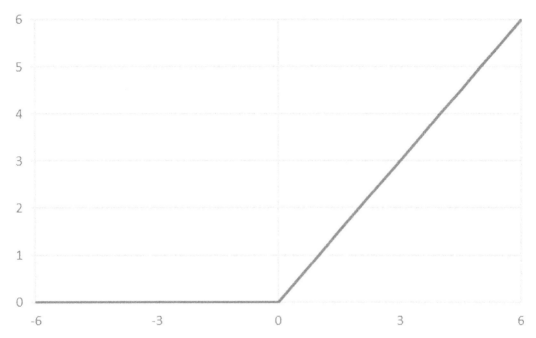

Figure 2.13: Graphical representation of the ReLU activation function

THE CALCULATION OF LOSS FUNCTIONS

Once forward propagation is complete, the next step in the training process is to calculate a loss function to estimate the error of the model by comparing how good or bad the prediction is in relation to the ground truth value. Considering this, the ideal value to be reached is 0, which would mean that there is no divergence between the two values.

This means that the goal in each iteration of the training process is to minimize the loss function by changing the parameters (weights and biases) that are used to perform the calculations during the forward pass.

Again, there are multiple loss functions to choose from. However, the most commonly used loss functions for regression and classification tasks are as follows:

- **Mean squared error (MSE)**: Widely used to measure the performance of regression models, the MSE function calculates the sum of the distance between the ground truth and the prediction values:

$$loss = \frac{1}{n} \sum_{i=1}^{n} (y_i - \hat{y}_i)^2$$

Figure 2.14: MSE loss function

Here, n refers to the number of samples, y_i is the ground truth values, and $\hat{y}_i$ is the predicted value.

- **Cross-entropy/multi-class cross-entropy**: This function is conventionally used for binary or multi-class classification models. It measures the divergence between two probability distributions; a large loss function will represent a large divergence. Hence, the objective here is to also minimize the loss function:

$$loss = -\frac{1}{n} \sum_{i=1}^{n} y_i * \log(\hat{y}_i) + (1 - y_i) * \log(1 - \hat{y}_i)$$

Figure 2.15: Cross-entropy loss function

Again, n refers to the number of samples. y_i and $\hat{y}_i$ are the ground truth and the predicted value, respectively.

BACKWARD PROPAGATION

The final step in the training process consists of going from right to left in the architecture of the network to calculate the partial derivatives (also known as gradients) of the loss function in respect to the weights and biases in each layer in order to update these parameters (weights and biases) so that in the next iteration step, the loss function is lower.

The final objective of the optimization algorithm is to find the global minima where the loss function has reached the least possible value, as shown in the following plot:

NOTE

A local minima refers to the smallest value within a section of the function domain. On the other hand, a global minima refers to the smallest value of the entire domain of the function.

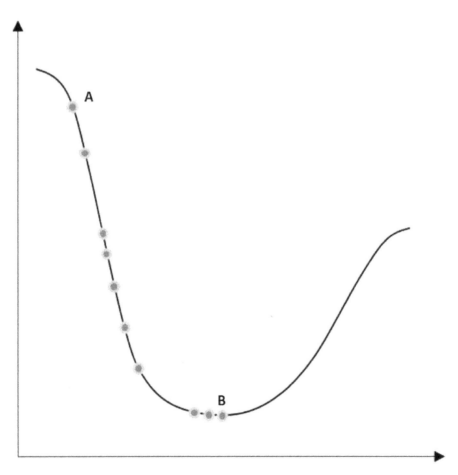

Figure 2.16: Loss function optimization through the iteration steps
in a two-dimensional space

Here, the dot furthest to the left, **A**, is the initial value of the loss function before any optimization. The dot furthest to the right, **B**, at the bottom of the curve, is the loss function after several iteration steps, where its value has been minimized. The process of going from one dot to another is called a **step**.

However, it is important to mention that the loss function is not always as smooth as the preceding one, which can introduce the risk of reaching a local minima during the optimization process.

This process is also called optimization, and there are different algorithms that vary in methodology to achieve the same objective. The most commonly used optimization algorithm will be explained next.

GRADIENT DESCENT

Gradient descent is the most widely used optimization algorithm among data scientists, and it is the basis of many other optimization algorithms. After the gradients for each neuron are calculated, the weights and biases are updated in the opposite direction of the gradient, which should be multiplied by a learning rate (used to control the size of the steps taken in each optimization), as shown in the following equations.

The learning rate is crucial during the training process as it prevents the update of the weights and biases from over/undershooting, which may prevent the model from reaching convergence or delay the training process, respectively.

The optimization of weights and biases in the gradient descent algorithm is as follows:

$$w' = w - \alpha * dw$$
$$b' = b - \alpha * db$$

Figure 2.17: Optimization of parameters in the gradient descent algorithm

Here, α refers to the learning rate, and *dw/db* represents the gradients of the weights or biases in a given neuron. The product of the two values is subtracted from the original value of the weight or bias in order to penalize the higher values, which are contributing to computing a large loss function.

An improved version of the gradient descent algorithm is called stochastic gradient descent, and it basically follows the same process, with the distinction that it takes the input data in random batches instead of in one chunk, which improves the training times while reaching outstanding performance. Moreover, this approach allows for the use of larger datasets because by using small batches of the dataset as inputs, we are no longer limited by computational resources.

ADVANTAGES AND DISADVANTAGES

The following is an explanation of the advantages and disadvantages of neural networks.

ADVANTAGES

Neural networks have become increasingly popular in the last few years for four main reasons:

- **Data**: Neural networks are widely known for their ability to capitalize on large amounts of data, and thanks to the advances in hardware and software, the collection and storage of massive databases is now possible. This has allowed neural networks to show their real potential as more data is fed into them.

- **Complex data problems**: As we explained previously, neural networks are excellent for solving complex data problems that cannot be tackled by other machine learning algorithms. This is mainly due to their ability to process large datasets and uncover complex patterns.

- **Computational power**: Advances in technology have also increased the computational power that's available these days, which is crucial for training neural network models that use millions of pieces of data.

- **Academic research**: Thanks to the preceding three points, a proliferation of academic research on this topic is available on the internet, which not only facilitates the immersion of new research each day, but also helps keep the algorithms and hardware/software requirements up to date.

DISADVANTAGES

Just because there are a lot of advantages to using a neural network does not mean that every data problem should be solved this way. This is a mistake that is commonly made. There is no one algorithm that will perform well for all data problems, and selecting the algorithm to use should depend on the resources available, as well as the data problem.

Although neural networks are thought to outperform almost any machine learning algorithm, it is crucial to consider their disadvantages as well so that you can weigh up what matters most for the data problem. Let's go through them now:

- **Black box**: This is one of the most commonly known disadvantages of neural networks. It basically means that how and why a neural network reached a certain output is unknown. For instance, when a neural network incorrectly predicts a cat picture as a dog, it is not possible to know what the cause of the error was.

- **Data requirements**: The vast amounts of data that they require to achieve optimal results can be equally an advantage and a disadvantage. Neural networks require more data than traditional machine learning algorithms, which can be the main reason to choose between them and other algorithms for some data problems. This becomes a greater issue when the task at hand is supervised, which means that the data needs to be labeled.

- **Training times**: Tied to the preceding disadvantage, the need for vast amounts of data also makes the training process last longer than traditional machine learning algorithms, which, in some cases, is not an option. Training times can be reduced through the use of GPUs, which speed up computation.

- **Computationally expensive**: Again, the training process of neural networks is computationally expensive. While one neural network could take weeks to converge, other machine learning algorithms could take hours or minutes to be trained. The amount of computational resources needed depends on the quantity of data at hand, as well as the complexity of the network; deeper neural networks take a longer time to train.

> **NOTE**
>
> There are a wide variety of neural network architectures. Three of the most commonly used ones will be explained in this chapter, along with their practical implementation in subsequent chapters. However, if you wish to learn about other architectures, visit http://www.asimovinstitute.org/neural-network-zoo/.

INTRODUCTION TO ARTIFICIAL NEURAL NETWORKS

Artificial neural networks (**ANNs**), also known as multi-layer perceptrons, are collections of multiple perceptrons. The connection between perceptrons occurs through layers. One layer can have as many perceptrons as desired, and they are all connected to all the other perceptrons in the preceding and subsequent layers.

Networks can have one or more layers. Networks with over four layers are considered to be deep neural networks and are commonly used to solve complex and abstract data problems.

ANNs are typically composed of three main elements, which were explained earlier, and can also be seen in the following image:

1. **Input layer**: This is the first layer of the network, conventionally located furthest to the left in the graphical representation of a network. It receives the input data before any calculation is performed and completes the first set of calculations. This is where the most generic patterns are uncovered.

    For supervised learning problems, the input data consists of a pair of features and targets. The job of the network is to uncover the correlation or dependency between the features and target.

2. **Hidden layers**: Next, the hidden layers can be found. A neural network can have many hidden layers, meaning there can be any number of layers between the input layer and the output layer. The more layers it has, the more complex data problems it can tackle, but it will also take longer to train. There are also neural network architectures that do not contain hidden layers at all, which is the case with single-layer networks.

    In each layer, a computation is performed based on the information that's received as input from the previous layer, which is then used to output a value that will become the input of the subsequent layer.

3. **Output layer**: This is the last layer of the network as is located at the far right of the graphical representation of the network. It receives data after the data has been processed by all the neurons in the network to make a final prediction.

The output layer can have one or more neurons. The former refers to models where the solution is binary, in the form of 0s or 1s. On the other hand, the latter case consists of models that output the probability of an instance belonging to each of the possible class labels (the possible values that the target variable has), meaning that the layer will have as many neurons as there are class labels:

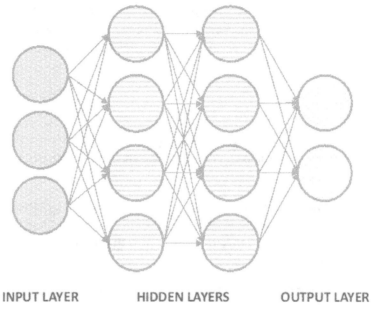

INPUT LAYER   HIDDEN LAYERS   OUTPUT LAYER

Figure 2.18: Architecture of a neural network with two hidden layers

INTRODUCTION TO CONVOLUTIONAL NEURAL NETWORKS

Convolutional Neural Networks (**CNNs**) are mostly used in the field of computer vision, where, in recent decades, machines have achieved levels of accuracy that surpass human ability.

CNNs create models that use subgroups of neurons to recognize different aspects of an image. These groups should be able to communicate with each other so that, together, they can form the complete image.

Considering this, the layers in the architecture of a CNN divide their recognition tasks. The first layers focus on trivial patterns, while the layers at the end of the network use that information to uncover more complex patterns.

For instance, when recognizing human faces in pictures, the first couple of layers focus on finding edges that separate one feature from another. Next, the subsequent layers emphasize certain features of the face, such as the nose. Finally, the last couple of layers use this information to put the entire face of the person together.

This idea of activating a group of neurons when certain features are encountered is achieved through the use of filters (kernels), which are one of the main building blocks of the architecture of CNNs. However, they are not the only elements present in the architecture, which is why a brief explanation of all the components of CNNs will be provided here:

> **NOTE**
>
> The concepts of padding and stride, which you might have heard of when using CNNs, will be explained in subsequent chapters of this book.

1. **Convolutional layers**: In these layers, a convolutional computation occurs between an image (represented as a matrix of pixels) and a filter. This computation produces a feature map as output that ultimately serves as input for the next layer.

   The computation takes a subsection of the image matrix of the same shape of the filter and performs a multiplication of the values. Then, the sum of the product is set as the output for that section of the image, as shown in the following diagram:

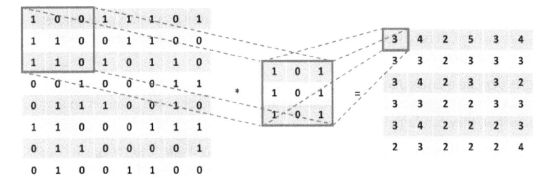

Figure 2.19: Convolution operation between the image and filter

Here, the matrix to the left is the input data, the matrix in the middle is the filter, and the matrix to the right is the output from the computation. The computation that occurred with the values highlighted by the boxes can be seen here:

$$1 * 1 + 1 * 1 + 1 * 1 + 0 * 0 + 1 * 0 + 1 * 0 + 0 * 1 + 0 * 1 + 0 * 1 = 3$$

Figure 2.20: Convolution of the first section of the image

This convolutional multiplication is done for all the subsections of the image. The following diagram shows another convolution step for the same example:

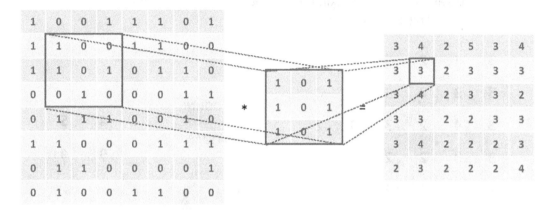

Figure 2.21: A further step in the convolution operation

One important notion of convolutional layers is that they are invariant in such a way that each filter will have a specific function, which does not vary during the training process. For instance, a filter in charge of detecting ears will only specialize in that function throughout the training process.

Moreover, a CNN will typically have several convolutional layers, considering that each of them will focus on identifying a particular feature or set of features of the image, depending on the filters that are used. Commonly, there is one pooling layer between two convolutional layers.

2. **Pooling layers**: Although convolutional layers are capable of extracting relevant features from images, their results can become enormous when analyzing complex geometrical shapes, which would make the training process impossible in terms of computational power, hence the invention of pooling layers.

These layers not only accomplish the goal of reducing the output of the convolutional layers, but also achieve the removal of any noise that's present in the features that have been extracted, which ultimately helps to increase the accuracy of the model.

There are two main types of pooling layers that can be applied, and the idea behind them is to detect the areas that express a stronger influence in the image so that the other areas can be overlooked.

Max pooling: This operation consists of taking a subsection of the matrix of a given size and taking the maximum number in that subsection as the output of the max pooling operation:

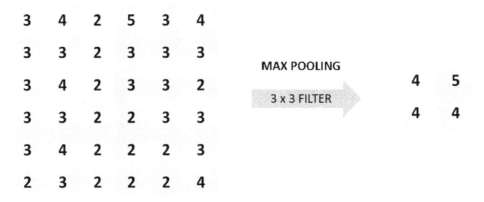

Figure 2.22: A max pooling operation

In the preceding diagram, by using a 3 x 3 max pooling filter, the result on the right is achieved. Here, the yellow section (top-left corner) has a maximum number of 4, while the orange section (top-right corner) has a maximum number of 5.

Average pooling: Similarly, the average pooling operation takes subsections of the matrix and takes the number that meets the rule as output, which, in this case, is the average of all the numbers in the subsection in question:

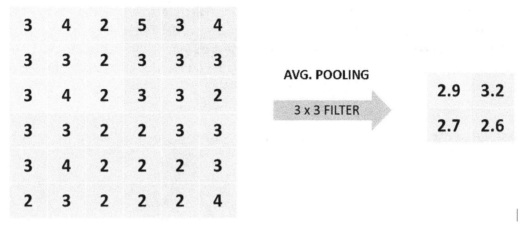

Figure 2.23: An average pooling operation

Here, using a 3 x 3 filter, we get 2.9, which is the average of all the numbers in the yellow section (top-left corner), while 3.2 is the average for the ones in the orange section (top-right corner).

3. **Fully connected layers**: Finally, considering that the network would be of no use if it was only capable of detecting a set of features without having the capability of classifying them into a class label, fully connected layers are used at the end of CNNs to take the features that were detected by the previous layer (known as the feature map) and output the probability of that group of features belonging to a class label, which is used to make the final prediction.

Like ANNs, fully connected layers use perceptrons to calculate an output based on a given input. Moreover, it is crucial to mention that CNNs typically have more than one fully connected layer at the end of the architecture.

By combining all of these concepts, the conventional architecture of CNNs is obtained. There can be as many layers of each type as desired, and each convolutional layer can have as many filters as desired (each for a particular task). Additionally, the pooling layer should have the same number of filters as the previous convolutional layer, as shown in the following image:

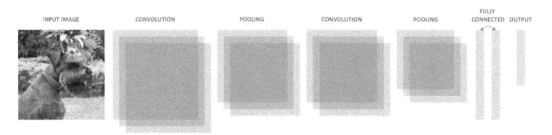

Figure 2.24: Diagram of the CNN architecture

INTRODUCTION TO RECURRENT NEURAL NETWORKS

The main limitation of the aforementioned neural networks (ANNs and CNNs) is that they learn only by considering the current event (the input that is being processed) without taking into account previous or subsequent events, which is inconvenient considering that we humans do not think that way. For instance, when reading a book, you can understand each sentence better by considering the context from the previous paragraph or more.

Due to this, and taking into account the fact that neural networks aim to optimize several processes that are traditionally done by humans, it is crucial to think of a network that's able to consider a sequence of inputs and outputs, hence the creation of **recurrent neural networks** (**RNNs**). They are a robust type of neural network that allow solutions to be found for complex data problems through the use of internal memory.

Simply put, these networks contain loops in them that allow for the information to remain in their memory for longer periods, even when a subsequent set of information is being processed. This means that a perceptron in an RNN not only passes over the output to the following perceptron, but it also retains a bit of information to itself, which can be useful for analyzing the next bit of information. This memory-keeping capability allows them to be very accurate in predicting what's coming next.

The learning process of an RNN, similar to other networks, tries to map the relationship between an input (x) and an output (y), with the difference being that these models also take into consideration the entire or partial history of previous inputs.

RNNs allow sequences of data to be processed in the form of a sequence of inputs, a sequence of outputs, or even both at the same time, as shown in the following diagram:

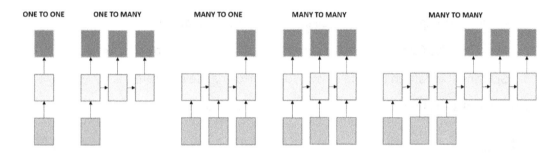

Figure 2.25: Sequence of data handled by RNNs

Here, each box is a matrix and the arrows represent a function that occurs. The bottom boxes are the inputs, the top boxes are the outputs, and the middle boxes represent the state of the RNN at that point, which holds the memory of the network.

From left to right, the preceding diagrams can be explained as follows:

1. A typical model that does not require an RNN to be solved. It has a fixed input and a fixed output. This can refer to image classification, for instance.

2. This model takes in an input and yields a sequence of outputs. Take, for instance, a model that receives an image as input; the output should be an image caption.

3. Contrary to the preceding model, this model takes a sequence of inputs and yields a single outcome. This type of architecture can be seen on sentiment analysis problems, where the input is the sentence to be analyzed and the output is the predicted sentiment behind the sentence.

4. The final two models take a sequence of inputs and return a sequence of outputs, with the difference being that the first one analyzes the inputs and generates the outputs at the same time; for example, when each frame of a video is being labeled individually. On the other hand, the second many-to-many model analyzes the entire set of inputs in order to generate the set of outputs. An example of this is language translation, where the entire sentence in one language needs to be understood before proceeding with the actual translation.

DATA PREPARATION

The first step in the development of any deep learning model – after gathering the data, of course – should be preparation of the data. This is crucial if we wish to understand the data at hand to outline the scope of the project correctly.

Many data scientists fail to do so, which results in models that perform poorly, and even models that are useless as they do not answer the data problem to begin with.

The process of preparing the data can be divided into three main tasks:

1. Understanding the data and dealing with any potential issues

2. Rescaling the features to make sure no bias is introduced by mistake

3. Splitting the data to be able to measure performance accurately

All three tasks will be further explained in the next section.

> **NOTE**
>
> All of the tasks we explained previously are pretty much the same when applying any machine learning algorithm, considering that they refer to the techniques that are required to prepare data beforehand.

DEALING WITH MESSY DATA

This task mainly consists of performing **exploratory data analysis** (**EDA**) to understand the data available, as well as to detect potential issues that may affect the development of the model.

The EDA process is useful as it helps the developer uncover information that's crucial to the definition of the course of action. This information is explained here:

1. **Quantity of data**: This refers both to the number of instances and the number of features. The former is crucial for determining whether it is necessary or even possible to solve the data problem using a neural network, or even a deep neural network, considering that such models require vast amounts of data to achieve high levels of accuracy. The latter, on the other hand, is useful for determining whether it would be a good practice to develop some feature selection methodologies beforehand in order to reduce the number of features, to simplify the model, and to eliminate any redundant information.

2. **The target feature**: For supervised models, data needs to be labeled. Considering this, it is highly important to select the target feature (the objective that we want to achieve by building the model) in order to assess whether the feature has many missing or outlier values. Additionally, this helps determine the objective of the development, which should be in line with the data that's available.

3. **Noisy data/outliers**: Noisy data refers to values that are visibly incorrect, for instance, a person who is 200 years old. On the other hand, outliers refer to values that, although they may be correct, are very far from the mean, for instance, a 10-year-old college student.

   There is not an exact science for detecting outliers, but there are some methodologies that are commonly accepted. Assuming a normally distributed dataset, one of the most popular ones is determining any value that is about 3-6 standard deviations away from the mean as an outlier.

   An equally valid approach to identifying outliers is to select those values at the 99th and 1st percentiles.

   It is very important to handle such values when they represent over 5% of the data for a feature because failing to do so may introduce bias to the model. The way to handle these values, as with any other machine learning algorithm, is to either delete the outlier values or assign new values using mean or regression imputation techniques.

4. **Missing values**: Similar to the aforementioned, a dataset with many missing values can introduce bias to the model, considering that different models will make different assumptions about those values. Again, when missing values represent over 5% of the values of a feature, they should be handled by eliminating or replacing them, again using the mean or regression imputation techniques.

5. **Qualitative features**: Finally, checking whether the dataset contains qualitative data is also a key step, considering that removing or encoding data may result in more accurate models.

   Additionally, in many research developments, several algorithms are tested on the same data in order to determine which one performs better, and some of these algorithms do not tolerate the use of qualitative data, as is the case with neural networks. This proves the importance of converting or encoding them to be able to feed all the algorithms the same data.

EXERCISE 2.02: DEALING WITH MESSY DATA

> **NOTE**
>
> All of the exercises in this chapter will be completed using the **Appliances energy prediction Dataset** sourced from the UC Irvine Machine Learning Repository, which was downloaded from https://archive.ics.uci.edu/ml/datasets/Appliances+energy+prediction. It can also be found in this book's GitHub repository: https://packt.live/34MBoSw
>
> The **Appliances energy prediction Dataset** contains 4.5 months of data related to temperature and humidity measures for different rooms in a low-energy building, with the objective of predicting the energy that's used by certain appliances.

In this exercise, we will use **pandas**, which is a popular Python package, to explore the data at hand and learn how to detect missing values, outliers, and qualitative values. Perform the following steps to complete this exercise:

> **NOTE**
>
> For the exercises and activities within this chapter, you will need to have Python 3.7, Jupyter 6.0, NumPy 1.17, and Pandas 0.25 installed on your local machine.

1. Open a Jupyter notebook to implement this exercise.

2. Import the pandas library:

```
import pandas as pd
```

3. Use pandas to read the CSV file containing the dataset we downloaded from the UC Irvine Machine Learning Repository site.

   Next, drop the column named **date** as we do not want to consider it for the following exercises:

```
data = pd.read_csv("energydata_complete.csv")
data = data.drop(columns=["date"])
```

Finally, print the head of the DataFrame:

```
data.head()
```

The output should look as follows:

	Appliances	lights	T1	RH_1	T2	RH_2	T3	RH_3	T4	RH_4	...	T9	RH_9
0	60	30	19.89	47.596667	19.2	44.790000	19.79	44.730000	19.000000	45.566667	...	17.033333	45.53
1	60	30	19.89	46.693333	19.2	44.722500	19.79	44.790000	19.000000	45.992500	...	17.066667	45.56
2	50	30	19.89	46.300000	19.2	44.626667	19.79	44.933333	18.926667	45.890000	...	17.000000	45.50
3	50	40	19.89	46.066667	19.2	44.590000	19.79	45.000000	18.890000	45.723333	...	17.000000	45.40
4	60	40	19.89	46.333333	19.2	44.530000	19.79	45.000000	18.890000	45.530000	...	17.000000	45.40

Figure 2.26: Top instances of the Appliances energy prediction dataset

4. Check for categorical features in your dataset:

```
cols = data.columns

num_cols = data._get_numeric_data().columns

list(set(cols) - set(num_cols))
```

The first line generates a list of all the columns in your dataset. Next, the columns that contain numeric values are stored in a variable as well. Finally, by subtracting the numeric columns from the entire list of columns, it is possible to obtain those that are not numeric.

The resulting list is empty, which indicates that there are no categorical features to deal with.

5. Use Python's **isnull()** and **sum()** functions to find out whether there are any missing values in each column of the dataset:

```
data.isnull().sum()
```

This command counts the number of null values in each column. For the dataset in use, there should not be any missing values, as can be seen here:

```
Appliances      0
lights          0
T1              0
RH_1            0
T2              0
RH_2            0
T3              0
RH_3            0
T4              0
RH_4            0
T5              0
RH_5            0
T6              0
RH_6            0
T7              0
RH_7            0
T8              0
RH_8            0
T9              0
RH_9            0
T_out           0
Press_mm_hg     0
RH_out          0
Windspeed       0
Visibility      0
Tdewpoint       0
rv1             0
rv2             0
dtype: int64
```

Figure 2.27: Missing values count

6. Use three standard deviations as the measure to detect any outliers for all the features in the dataset:

```
outliers = {}
for i in range(data.shape[1]):
    min_t = data[data.columns[i]].mean() \
            - (3 * data[data.columns[i]].std())
    max_t = data[data.columns[i]].mean() \
            + (3 * data[data.columns[i]].std())

    count = 0
    for j in data[data.columns[i]]:
        if j < min_t or j > max_t:
            count += 1

    percentage = count / data.shape[0]
    outliers[data.columns[i]] = "%.3f" % percentage

outliers
```

The preceding code snippet performs a **for** loop through the columns in the dataset in order to evaluate the presence of outliers in each of them. It continues to calculate the minimum and maximum thresholds so that it can count the number of instances that fall outside the range between the thresholds.

Finally, it calculates the percentage of outliers (that is, the number of outliers divided by the total number of instances) in order to output a dictionary that displays this percentage for each column.

By printing the resulting dictionary (**outliers**), it is possible to display a list of all the features (columns) in the dataset, along with the percentage of outliers. According to the result, it is possible to conclude that there is no need to deal with the outlier values, considering that they account for less than 5% of the data, as can be seen in the following screenshot:

NOTE

Note that Jupyter Notebooks can print the value of a variable without the need for the print function whenever the variable is placed at the end of a cell in the notebook. In any other programming platform or any other scenario, make sure to use the print function.

For instance, an equivalent way (and the best practice) to print the resulting dictionary containing the outliers would be to use the print statement, as follows: **print(outliers)**. This way, the code will have the same output when run in a different programming platform.

```
{'Appliances': '0.027',
 'lights': '0.033',
 'T1': '0.001',
 'RH_1': '0.006',
 'T2': '0.010',
 'RH_2': '0.007',
 'T3': '0.003',
 'RH_3': '0.001',
 'T4': '0.000',
 'RH_4': '0.000',
 'T5': '0.001',
 'RH_5': '0.029',
 'T6': '0.005',
 'RH_6': '0.000',
 'T7': '0.000',
 'RH_7': '0.001',
 'T8': '0.000',
 'RH_8': '0.000',
 'T9': '0.000',
 'RH_9': '0.000',
 'T_out': '0.005',
 'Press_mm_hg': '0.005',
 'RH_out': '0.008',
 'Windspeed': '0.005',
 'Visibility': '0.002',
 'Tdewpoint': '0.000',
 'rv1': '0.000',
 'rv2': '0.000'}
```

Figure 2.28: Outlier participation in each feature

> **NOTE**
>
> To access the source code for this specific section, please refer to
> https://packt.live/2CYEglp.
>
> You can also run this example online at https://packt.live/3ePAg4G. You must
> execute the entire Notebook in order to get the desired result.

You have successfully explored the dataset and dealt with potential issues.

DATA RESCALING

Although data does not need to be rescaled to be fed to an algorithm for training, it is an important step if you wish to improve a model's accuracy. This is basically because having different scales for each feature may result in the model assuming that a given feature is more important than others as it has higher numerical values.

Take, for instance, two features, one measuring the number of children a person has and another stating the age of the person. Even though the age feature may have higher numerical values, in a study for recommending schools, the number of children feature may be more important.

Considering this, if all the features are scaled equally, the model can actually give higher weights to those features that matter the most in respect to the target feature, and not the numerical values that they have. Moreover, it can also help accelerate the training process by removing the need for the model to learn from the invariance of the data.

There are two main rescaling methodologies that are popular among data scientists, and although there is no rule for selecting one or the other, it is important to highlight that they are to be used individually (one or the other).

A brief explanation of both of these methodologies can be found here:

- **Normalization**: This consists of rescaling the values so that all the values of all the features are between zero and one. This is done using the following equation:

$$x_{norm} = \frac{x_i - min(x)}{max(x) - min(x)}$$

Figure 2.29: Data normalization

- **Standardization**: In contrast, this rescaling methodology converts all the values so that their mean is 0 and their standard deviation is equal to 1. This is done using the following equation:

$$x_{standardization} = \frac{x_i - mean(x)}{std(x)}$$

Figure 2.30: Data standardization

EXERCISE 2.03: RESCALING DATA

In this exercise, we will rescale the data from the previous exercise. Perform the following steps to do so:

> **NOTE**
>
> Use the same Jupyter notebook that you used in the previous exercise.

1. Separate the features from the target. We are only doing this to rescale the features data:

```
X = data.iloc[:, 1:]
Y = data.iloc[:, 0]
```

The preceding code snippet takes the data and uses slicing to separate the features from the target.

2. Rescale the features data by using the normalization methodology. Display the head (that is, the top five instances) of the resulting DataFrame to verify the result:

```
X = (X - X.min()) / (X.max() - X.min())
X.head()
```

The output should look as follows:

	lights	T1	RH_1	T2	RH_2	T3	RH_3	T4	RH_4	T5	...	T9
0	0.428571	0.32735	0.566187	0.225345	0.684038	0.215188	0.746066	0.351351	0.764262	0.175506	...	0.223032
1	0.428571	0.32735	0.541326	0.225345	0.682140	0.215188	0.748871	0.351351	0.782437	0.175506	...	0.226500
2	0.428571	0.32735	0.530502	0.225345	0.679445	0.215188	0.755569	0.344745	0.778062	0.175506	...	0.219563
3	0.571429	0.32735	0.524080	0.225345	0.678414	0.215188	0.758685	0.341441	0.770949	0.175506	...	0.219563
4	0.571429	0.32735	0.531419	0.225345	0.676727	0.215188	0.758685	0.341441	0.762697	0.178691	...	0.219563

Figure 2.31: Top instances of the normalized Appliances energy prediction dataset

> **NOTE**
>
> To access the source code for this specific section, please refer to https://packt.live/2ZojumJ.
>
> You can also run this example online at https://packt.live/2NLVgxq.
> You must execute the entire Notebook in order to get the desired result.

You have successfully rescaled a dataset.

SPLITTING THE DATA

The purpose of splitting the dataset into three subsets is so that the model can be trained, fine-tuned, and measured appropriately, without the introduction of bias. Here is an explanation of each set:

- **Training set**: As its name suggests, this set is fed to the neural network to be trained. For supervised learning, it consists of the features and the target values. This is typically the largest set out of the three, considering that neural networks require large amounts of data to be trained, as we mentioned previously.

- **Validation set (dev set)**: This set is used mainly to measure the performance of the model in order to make adjustments to the hyperparameters to improve performance. This fine-tuning process is done so that we can configure the hyperparameters that achieve the best results.

  Although the model is not trained on this data, it indirectly has an effect on it, which is why the final measure of performance should not be done on it as it may be a biased measure.

- **Testing set**: This set does not have an effect on the model, which is why it is used to perform a final evaluation of the model on unseen data, which becomes a guideline of how well the model will perform on future datasets.

There is no actual science on the perfect ratio for splitting data into the three sets mentioned, considering that every data problem is different and developing deep learning solutions usually requires a trial-and-error methodology. Nevertheless, it is widely known that larger datasets (hundreds of thousands and millions of instances) should have a split ratio of 98:1:1 for each set, considering that it is crucial to use as much data as possible for the training set. For a smaller dataset, the conventional split ratio is 60:20:20.

EXERCISE 2.04: SPLITTING A DATASET

In this exercise, we will split the dataset from the previous exercise into three subsets. For the purpose of learning, we will explore two different approaches. First, the dataset will be split using indexing. Next, scikit-learn's **train_test_split()** function will be used for the same purpose, thereby achieving the same result with both approaches. Perform the following steps to complete this exercise:

> **NOTE**
>
> Use the same Jupyter notebook that you used in the previous exercise.

1. Print the shape of the dataset in order to determine the split ratio to be used:

    ```
    X.shape
    ```

    The output from this operation should be **(19735, 27)**. This means that it is possible to use a split ratio of 60:20:20 for the training, validation, and test sets.

2. Get the value that you will use as the upper bound of the training and validation sets. This will be used to split the dataset using indexing:

    ```
    train_end = int(len(X) * 0.6)
    dev_end = int(len(X) * 0.8)
    ```

    The preceding code determines the index of the instances that will be used to divide the dataset through slicing.

3. Shuffle the dataset:

```
X_shuffle = X.sample(frac=1, random_state=0)
Y_shuffle = Y.sample(frac=1, random_state=0)
```

Using the pandas **sample** function, it is possible to shuffle the elements in the features and target matrices. By setting **frac** to 1, we ensure that all the instances are shuffled and returned in the output from the function. Using the **random_state** argument, we ensure that both datasets are shuffled equally.

4. Use indexing to split the shuffled dataset into the three sets for both the features and the target data:

```
x_train = X_shuffle.iloc[:train_end,:]
y_train = Y_shuffle.iloc[:train_end]
x_dev = X_shuffle.iloc[train_end:dev_end,:]
y_dev = Y_shuffle.iloc[train_end:dev_end]
x_test = X_shuffle.iloc[dev_end:,:]
y_test = Y_shuffle.iloc[dev_end:]
```

5. Print the shapes of all three sets:

```
print(x_train.shape, y_train.shape)
print(x_dev.shape, y_dev.shape)
print(x_test.shape, y_test.shape)
```

The result of the preceding operation should be as follows:

```
(11841, 27) (11841,)
(3947, 27) (3947,)
(3947, 27) (3947,)
```

6. Import the **train_test_split()** function from scikit-learn's **model_selection** module:

```
from sklearn.model_selection import train_test_split
```

> **NOTE**
>
> Although the different packages and libraries are being imported as they are needed for practical learning purposes, it is always good practice to import them at the beginning of your code.

7. Split the shuffled dataset:

```
x_new, x_test_2, \
y_new, y_test_2 = train_test_split(X_shuffle, Y_shuffle, \
                                   test_size=0.2, \
                                   random_state=0)

dev_per = x_test_2.shape[0]/x_new.shape[0]

x_train_2, x_dev_2, \
y_train_2, y_dev_2 = train_test_split(x_new, y_new, \
                                      test_size=dev_per, \
                                      random_state=0)
```

The first line of code performs an initial split. The function takes the following as arguments:

X_shuffle, **Y_shuffle**: The datasets to be split, that is, the features dataset, as well as the target dataset (also known as X and Y)

test_size: The percentage of instances to be contained in the testing set

random_state: Used to ensure the reproducibility of the results

The result from this line of code is the division of each of the datasets (X and Y) into two subsets.

To create an additional set (the validation set), we will perform a second split. The second line of the preceding code is in charge of determining the **test_size** to be used for the second split so that both the testing and validation sets have the same shape.

Finally, the last line of code performs the second split using the value that was calculated previously as the **test_size**.

8. Print the shape of all three sets:

```
print(x_train_2.shape, y_train_2.shape)
print(x_dev_2.shape, y_dev_2.shape)
print(x_test_2.shape, y_test_2.shape)
```

The result from the preceding operation should be as follows:

```
(11841, 27) (11841,)
(3947, 27) (3947,)
(3947, 27) (3947,)
```

As we can see, the resulting sets from both approaches have the same shapes. Using one approach or the other is a matter of preference.

> **NOTE**
>
> To access the source code for this specific section, please refer to https://packt.live/2VxvroW.
>
> You can also run this example online at https://packt.live/3gcm5H8. You must execute the entire Notebook in order to get the desired result.

You have successfully split the dataset into three subsets.

DISADVANTAGES OF FAILING TO PREPARE YOUR DATA

Although the process of preparing the dataset is time-consuming and may be tiring when dealing with large datasets, the disadvantages of failing to do so are even more inconvenient:

- **Longer training times**: Data containing noise, missing values, and redundant or irrelevant columns takes considerably longer to train and, in most cases, this delay in time is even longer than the time it takes to prepare the data. For instance, during data preparation, it may be determined that five columns are irrelevant for the purpose of the study, which may reduce the dataset considerably, and hence reduce the training times considerably.

- **Introduction of bias**: Uncleaned data usually contains errors or missing values that can deviate the model from the truth. For instance, missing values can cause the model to make inferences that are not true, which, in turn, creates a model that does not represent the data.

- **Avoid generalization**: Outliers and noisy values prevent the model from making generalizations of the data, which is crucial for building a model that represents the current training data, as well as future unseen data. For example, a dataset containing a variable for age that contains entries of people who are over 100 years old may result in a model that accounts for those users who, in reality, represent a very small portion of the population.

ACTIVITY 2.01: PERFORMING DATA PREPARATION

In this activity, we will prepare a dataset containing a list of songs, each with several attributes that help determine the year they were released. This data preparation step is crucial for the next activity in this chapter. Let's look at the following scenario.

You work at a music record company and your boss wants to uncover the details that characterize records from different time periods, which is why they have put together a dataset that contains data on 515,345 records, with release years ranging from 1922 to 2011. They have tasked you with preparing the dataset so that it is ready to be fed to a neural network. Perform the following steps to complete this activity:

> **NOTE**
>
> To download the dataset for this activity, visit the following UC Irvine Machine Learning Repository URL: https://archive.ics.uci.edu/ml/datasets/YearPredictionMSD.
>
> Citation: Dua, D. and Graff, C. (2019). UCI Machine Learning Repository [http://archive.ics.uci.edu/ml]. Irvine, CA: University of California, School of Information and Computer Science.
>
> It is also available at this book's GitHub repository: https://packt.live/38kZzZR

1. Import the required libraries.

2. Using pandas, load the `.csv` file.

3. Verify whether any qualitative data is present in the dataset.

4. Check for missing values.

   You can also add an additional **sum()** function to get the sum of missing values in the entire dataset, without discriminating by column.

5. Check for outliers.

6. Separate the features from the target data.

7. Rescale the data using the standardization methodology.

8. Split the data into three sets: training, validation, and test. Use whichever approach you prefer.

> **NOTE**
>
> The solution to this activity can be found on page 239.

BUILDING A DEEP NEURAL NETWORK

Building a neural network, in general terms, can be achieved either on a very simple level using libraries such as scikit-learn (not suitable for deep learning), which perform all the math for you without much flexibility, or on a very complex level by coding every single step of the training process from scratch, or by using a more robust framework, which allows great flexibility.

PyTorch was built considering the input of many developers in the field and has the advantage of allowing both approximations in the same place. As we mentioned previously, it has a neural network module that was built to allow easy predefined implementations of simple architectures using the sequential container, while at the same time allowing for the creation of custom modules that introduce flexibility to the process of building very complex architectures.

In this section, we will discuss the use of the sequential container for developing deep neural networks in order to demystify their complexity. Nevertheless, in later sections of this book, we will move on and explore more complex and abstract applications, which can also be achieved with very little effort.

As we mentioned previously, the sequential container is a module that was built to contain sequences of modules that follow an order. Each of the modules it contains will apply some computation to a given input to arrive at an outcome.

Some of the most popular modules (layers) that can be used inside the sequential container to develop regular classification models are explained here:

> **NOTE**
>
> The modules that are used for other types of architectures, such as CNNs and RNNs, will be explained in subsequent chapters.

- **Linear layer**: This applies a linear transformation to the input data while keeping internal tensors to hold the weights and biases. It receives the size of the input sample (the number of features of the dataset or the number of outputs from the previous layer), the size of the output sample (the number of units in the current layer, which will be the number of outputs), and whether to use a tensor of biases during the training process (which is set to **True** by default) as arguments.

- **Activation functions**: They receive the output from the linear layer as input in order to break the linearity. There are several activation functions, as explained previously, that can be added to the sequential container. The most commonly used ones are explained here:

  ReLU: This applies the rectified linear unit function to the tensor containing the input data. The only argument it takes in is whether the operation should be done in-situ, which is set to **False** by default.

  Tanh: This applies the element-wise tanh function to the tensor containing the input data. It does not take any arguments.

  Sigmoid: This applies the previously explained sigmoid function to the tensor containing the input data. It does not take any arguments.

  Softmax: This applies the softmax function to an n-dimensional tensor containing the input data. The output is rescaled so that the elements of the tensor lie in a range between zero and one, and sum to one. It takes the dimension along which the softmax function should be computed as an argument.

- **Dropout layer**: This module randomly zeroes some of the elements of the input tensor, according to a set probability. It takes the probability to use for the random selection, as well as whether the operation should be done in-situ, which is set to **False** by default, as arguments. This technique is commonly used for dealing with overfitted models, which will be explained in more detail later.

- **Normalization layer**: There are different methodologies that can be used to add a normalization layer in the sequential container. Some of them include **BatchNorm1d**, **BatchNorm2d**, and **BatchNorm3d**. The idea behind this is to normalize the output from the previous layer, which ultimately achieves similar accuracy levels at lower training times.

EXERCISE 2.05: BUILDING A DEEP NEURAL NETWORK USING PYTORCH

In this exercise, we will use the PyTorch library to define the architecture of a deep neural network of four layers, which will then be trained with the dataset we prepared in the previous exercises. Perform the following steps to do so:

> **NOTE**
>
> Use the same Jupyter notebook that you used in the previous exercise.

1. Import the PyTorch library, called **torch**, as well as the **nn** module from PyTorch:

```
import torch
import torch.nn as nn
```

> **NOTE**
>
> **torch.manual_seed(0)** is used in this exercise in order to ensure the reproducibility of the results obtained in this book's GitHub repository.

2. Separate the feature columns from the target for each of the sets we created in the previous exercise. Additionally, convert the final DataFrames into tensors:

```
x_train = torch.tensor(x_train.values).float()
y_train = torch.tensor(y_train.values).float()

x_dev = torch.tensor(x_dev.values).float()
y_dev = torch.tensor(y_dev.values).float()

x_test = torch.tensor(x_test.values).float()
y_test = torch.tensor(y_test.values).float()
```

3. Define the network architecture using the **sequential()** container. Make sure to create a four-layer network. Use ReLU activation functions for the first three layers and leave the last layer without an activation function, considering the fact that we are dealing with a regression problem.

The number of units for each layer should be 100, 50, 25, and 1:

```
model = nn.Sequential(nn.Linear(x_train.shape[1], 100), \
                      nn.ReLU(), \
                      nn.Linear(100, 50), \
                      nn.ReLU(), \
                      nn.Linear(50, 25), \
                      nn.ReLU(), \
                      nn.Linear(25, 1))
```

4. Define the loss function as the MSE:

```
loss_function = torch.nn.MSELoss()
```

5. Define the optimizer algorithm as the Adam optimizer:

```
optimizer = torch.optim.Adam(model.parameters(), lr=0.001)
```

6. Use a **for** loop to train the network over the training data for 1,000 iteration steps:

```
for i in range(1000):
    y_pred = model(x_train).squeeze()
    loss = loss_function(y_pred, y_train)
    optimizer.zero_grad()
    loss.backward()
    optimizer.step()

    if i%100 == 0:
        print(i, loss.item())
```

> **NOTE**
>
> The **squeeze()** function is used to remove the additional dimension of **y_pred**, which is converted from being of size [3000,1] to [3000].
>
> This is crucial considering that **y_train** is one-dimensional and both tensors need to have the same dimensions to be fed to the loss function.

Running the preceding snippet will yield an output similar to the following:

```
0  20351.2734375
100  10840.4326171875
200  10686.625
300  10449.0927734375
400  10123.974609375
500  9767.02734375
600  9480.8447265625
700  9206.1142578125
800  8968.525390625
900  8721.1806640625
```

Figure 2.32: Loss value for different iteration steps

As can be seen, the loss value continually decreases over time.

7. To test the model, perform a prediction on the first instance of the testing set and compare it with the ground truth (target value):

```
pred = model(x_test[0])
print("Ground truth:", y_test[0].item(), \
      "Prediction:",pred.item())
```

The output should look similar to the following:

```
Ground truth: 60.0 Prediction:  69.5818099975586
```

As you can see, the ground truth value (**60**) is fairly close to the predicted one (**69.58**).

> **NOTE**
>
> To access the source code for this specific section, please refer to https://packt.live/2NJsQUz.
>
> You can also run this example online at https://packt.live/38nrnNh.
> You must execute the entire Notebook in order to get the desired result.

You have successfully created and trained a deep neural network to solve a regression problem.

ACTIVITY 2.02: DEVELOPING A DEEP LEARNING SOLUTION FOR A REGRESSION PROBLEM

In this activity, we will create and train a neural network to solve the regression problem we mentioned in the previous activity. Let's look at the scenario.

You continue to work at the music record company and, after seeing the great job you did preparing the dataset, your boss has trusted you with the task of defining the network's architecture, as well as training it with the prepared dataset. Perform the following steps to complete this activity:

> **NOTE**
>
> Use the same Jupyter notebook that you used in the previous activity.

1. Import the required libraries.

2. Split the features from the targets for all three sets of data that we created in the previous activity. Convert the DataFrames into tensors.

3. Define the architecture of the network. Feel free to try different combinations of the number of layers and the number of units per layer.

4. Define the loss function and the optimizer algorithm.

5. Use a **for** loop to train the network for 3,000 iteration steps.

6. Test your model by performing a prediction on the first instance of the test set and comparing it with the ground truth.

Your output should look similar to this:

```
Ground truth: 1995.0 Prediction: 1998.0279541015625
```

> **NOTE**
>
> The solution to this activity can be found on page 242.

SUMMARY

The theory that gave birth to neural networks was developed decades ago by Frank Rosenblatt. It started with the definition of the perceptron, a unit inspired by the human neuron, that takes data as input to perform a transformation on it. The theory behind the perceptron consisted of assigning weights to input data to perform a calculation so that the end result would be either one thing or the other, depending on the outcome.

The most widely known form of neural networks is the one that's created from a succession of perceptrons, stacked together in layers, where the output from one column of perceptrons (layer) is the input for the following one.

The typical learning process for a neural network was explained. Here, there are three main processes to consider: forward propagation, the calculation of the loss function, and backpropagation.

The end goal of this procedure is to minimize the loss function by updating the weights and biases that accompany each of the input values in every neuron of the network. This is achieved through an iterative process that can take minutes, hours, or even weeks, depending on the nature of the data problem.

The main architecture of the three main types of neural networks was also discussed: the artificial neural network, the convolutional neural network, and the recurrent neural network. The first is used to solve traditional classification or regression problems, the second one is widely popular for its capacity to solve computer vision problems (for instance, image classification), and the third one is capable of processing data in sequence, which is useful for tasks such as language translation.

In the next chapter, the main differences between solving regression and a classification data problem will be discussed. You will also learn how to solve a classification data problem, as well as how to improve its performance and how to deploy the model.

3

A CLASSIFICATION PROBLEM USING DNN

OVERVIEW

In this chapter, we'll look at a real-life example in the banking industry in order to solve a classification data problem. You will learn how to make use of PyTorch's custom modules to define the network architecture and train the model. You will also explore the concept of error analysis to improve a model's performance. Finally, you will study the different approaches to deploying a model in order to make use of it in the future. By the end of this chapter, you will have a firm understanding of this process so that you can develop an end-to-end solution to a classification data problem using a **deep neural network** (**DNN**) in PyTorch.

INTRODUCTION

In the previous chapter, we learned about the building blocks of DNNs and reviewed the characteristics of the three most common architectures. Additionally, we learned how to solve a regression problem using a DNN.

In this chapter, we will use DNNs to solve a classification task, where the objective is to predict an outcome from a series of options.

One field that makes use of such models is banking. This is mainly due to their need to predict future behavior based on demographic data, alongside the main objective of ensuring profitability in the long term. Some of the uses in the banking sector include the evaluation of loan applications, credit card approval, the prediction of stock market prices, and the detection of fraud by analyzing behavior.

This chapter will focus on solving a classification banking problem using a deep **artificial neural network** (**ANN**), following all the steps required to arrive at an effective model: data exploration, data preparation, architecture definition and model training, model fine-tuning, error analysis, and, finally, deploying the final model.

> **NOTE**
>
> All the code present in this chapter can be found at:
> https://packt.live/38qLadV.

PROBLEM DEFINITION

Defining the problem is as important as building your model or improving accuracy. This is because, while you may be able to use the most powerful algorithm and use the most advanced methodologies to improve its results, this may prove pointless if you are solving the wrong problem or using the wrong data.

It is crucial to learn how to think deeply to understand what can and cannot be done, and how what can be done can be accomplished. This is especially important considering that when we are learning to apply machine learning or deep learning algorithms, the problems presented in most courses are always clearly defined, and there is no need for further analysis other than training the model and improving its performance. On the other hand, in real life, problems are often confusing, and data is often messy.

In this section, you will learn about some of the best practices for defining your problem based on the needs of your organization and on the data that you have at hand.

To do this, the steps that you need to follow are as follows:

1. Understand the what, why, and how of the problem.

2. Analyze the data at hand to determine some of the key parameters of our model, such as the type of learning task to be performed, the necessary preparation, and the definition of the performance metric.

3. Perform data preparation to reduce the probability of introducing bias to the model.

DEEP LEARNING IN BANKING

Banks and financial entities deal with huge amounts of information every day, which is required so that they can make crucial decisions that not only impact the future of their own organization but also those of the millions of individuals who trust them.

These decisions are made every second and, back in the 1990s, people in the banking sector used to rely on expert systems to code rule-based programs; that is, programs based on the knowledge of human experts to make up a set of rules to follow. Not surprisingly, such programs fell short as they required all the information or possible scenarios to be programmed upfront, which made them inefficient for dealing with uncertainty and highly changing markets.

As technology improved, the banking sector has been leading the transition to more specialized systems that make use of statistical models to help make such decisions. Moreover, as banks need to consider both their own profitability as well as that of their clients, they are considered notable among the many industries that constantly keep up with technological improvements.

Nowadays, along with the healthcare market, the banking and financial industries are driving the market of neural networks. This is mostly due to the capability of neural networks to deal with uncertainty by using vast amounts of previous data to predict future behavior. This is something that human expert knowledge-based systems are unable to achieve, considering that the human brain is not capable of analyzing such large amounts of data.

Some of the fields in the banking and financial services in which deep learning is being used are presented and explained in brief here:

- **Loan application evaluation**: Banks issue loans to customers based on different factors, including demographical information, credit history, and so on. Their main objective in this process is to minimize the number of customers who will default on loan payments (minimize the failure rate), thereby maximizing the returns obtained through the loans that have been issued.

  Neural networks are used to help make the decision of whether to grant a loan. They are usually trained using data from previous loan applicants who failed to pay loans back, as well as those who paid the loans on time. Once a model is created, the idea is to input the data of a new applicant into the model in order to get a prediction of whether they will pay the loan back, considering that the focus of the model should be to reduce the number of false positives (customers who the model predicted would default on the loan, but actually did not).

  It is known in the industry that the failure rate of neural networks is lower than traditional methodologies that rely on human expertise.

- **Detection of fraud**: Fraud detection is crucial for banks and financial providers, considering the advancements of technology that, in spite of making our lives easier, also leave us exposed to greater financial risks.

  Neural networks are used in this domain, specifically CNNs, for character and image recognition to detect hidden and abstract patterns in images of characters, in order to determine whether the user is being subjected to fraud.

- **Credit card customer selection**: To remain profitable in the long term, credit card providers need to find the right customers. For instance, approving a credit card to a customer with limited credit card needs (that is, a customer who will not use it) will result in low credit card revenues.

  On the other hand, credit card issuers are also interested in predicting whether a client will default on their next payment. This will help card issuers know in advance the quantity of funds that will be defaulted so that they can make preparations in order to stay profitable.

Networks are trained using the historical data of customers holding one or several credit cards. The objective is to create models that are capable of determining whether a new customer will make good use of the credit card so that the revenue will supersede cost—along with models that are able to predict payment behavior.

> **NOTE**
>
> The remainder of this chapter will focus on solving a data problem to do with credit card use. To download the dataset that will be used, go to http://archive.ics.uci.edu/ml/datasets/default+of+credit+card+clients, click on the `Data Folder` link, and download the `.xls` file. The dataset is also available in this book's GitHub repository, at https://packt.live/38qLadV.

EXPLORING THE DATASET

In the following sections, we will focus on solving a classification task related to credit card payments using the **Default of Credit Card Clients** (**DCCC**) dataset, which was previously downloaded from the UC Irvine Repository site.

The main idea behind this section is to clearly state the what, why, and how of the data problem, which will help determine the purpose of the study and the evaluation metric. Additionally, we will analyze the data at hand in detail in order to identify some of the steps required during the preparation of the data (for instance, converting qualitative features into their numerical representations).

First of all, let's define the what, why, and how. Take into consideration that this should be done to identify the real needs of the organization:

What: Build a model that is able to determine whether a client will default on the upcoming payment.

Why: To be able to predict the amount (in money) of payments to be received in the following month. This will help companies determine the spending strategies for that month and allow them to define the actions to be carried out with each customer, both to ensure future payments from those who will pay their bills and to improve the probabilities of payments of those clients who will default.

How: Use historical data containing demographical information, credit histories, and previous bill statements of clients who have and have not defaulted on their payments to train a model. After being trained over the input data, this model should be able to determine whether a client is likely to default the next payment.

Considering this, it seems that the target feature should be one stating whether a client will default on the next payment, which entails a binary outcome (yes/no). This means that the learning task to be developed is a classification task, and therefore the loss function should be one that's capable of measuring differences for such a type of learning (for instance, the cross-entropy function, as explained in the previous chapter).

Once the problem is well defined, you need to determine the priorities of the final model. This means determining whether all the output classes are equally important. For instance, a model measuring whether a lung mass is malignant should focus primarily on minimizing `false negatives` (patients who the model predicted as not having a malignant mass, but the mass was actually malignant), while a model built to recognize handwritten characters should not focus on one particular character, but rather maximize its performance in recognizing all characters equally.

Considering this, as well as the explanation in the why statement, the priority of the model for the `Default of Credit Card Clients` dataset should be to maximize the overall performance of the model, without prioritizing any of the class labels. This is mainly because the why statement declares that the main purpose of the study should be to have a better idea of the money that the bank will receive, as well as to perform certain actions regarding clients who are likely to default on a payment (for example, offering to split the debt into smaller payments), and different actions for those who will not default on it (for instance, offering benefits as a reward for being a well-behaved client).

According to this, the performance metric to be used in this case study is **accuracy**, which focuses on maximizing the **correctly classified instances**. This refers to the ratio of the correctly classified instances of any of the class labels to the total number of instances.

The following table contains a brief explanation of each of the features present in the dataset, which can help determine their relevance to the purpose of the study, as well as identify some of the preparation tasks that need to be performed:

Name of the feature	Type	Description	Relevance
ID	Quantitative. Nominal.	ID of the customer.	Irrelevant. An identification number does not provide useful information to the study.
LIMIT_BAL	Quantitative. Nominal.	Amount of credit in NT (New Taiwan) dollars. Includes individual and family credit.	Relevant.
SEX	Quantitative. Nominal.	Gender of the customer.	Irrelevant. The gender is irrelevant in determining whether a person will pay the next bill.
EDUCATION	Quantitative. Ordinal.	Level of education of the customer. 1 = grad school 2 = university 3 = high school 4 = others 5 = unknown 6 = unknown	Relevant. This feature may be linked to the earnings of a person.
MARRIAGE	Quantitative. Nominal.	Marital status of the customer. 1 = married 2 = single 3 = others	Relevant. This feature may be linked to the total earnings of the household.
AGE	Quantitative. Ordinal.	Age in years of the customer.	Relevant. May also be related to the earnings of the person.
PAY_0	Quantitative. Ordinal.	Payment status in September 2005. -2 = pay duly -1 = payment delay 1 month 0 = payment delay 2 months ... 7 = payment delay 8 months 8 = payment delay 9 months or over	Relevant. Refers to previous payment behavior of the customer.
PAY_2	Quantitative. Ordinal.	Payment status in August 2005. (Same scale as PAY_0)	Relevant.
PAY_3	Quantitative. Ordinal.	Payment status in July 2005. (Same scale as PAY_0)	Relevant.

Figure 3.1: A description of features from the DCCC dataset

Name of the feature	Type	Description	Relevance
PAY_4	Quantitative. Ordinal.	Payment status in June 2005. (Same scale as PAY_0)	Relevant.
PAY_5	Quantitative. Ordinal.	Payment status in May 2005. (Same scale as PAY_0)	Relevant.
PAY_6	Quantitative. Ordinal.	Payment status in April 2005. (Same scale as PAY_0)	Relevant.
BILL_AMT1	Quantitative. Nominal.	Amount of bill statement in September 2005 in NT dollars.	Relevant. Refers to previous spending behavior.
BILL_AMT2	Quantitative. Nominal.	Amount of bill statement in August 2005 in NT dollars.	Relevant.
BILL_AMT3	Quantitative. Nominal.	Amount of bill statement in July 2005 in NT dollars.	Relevant.
BILL_AMT4	Quantitative. Nominal.	Amount of bill statement in June 2005 in NT dollars.	Relevant.
BILL_AMT5	Quantitative. Nominal.	Amount of bill statement in May 2005 in NT dollars.	Relevant.
BILL_AMT6	Quantitative. Nominal.	Amount of bill statement in April 2005 in NT dollars.	Relevant.
PAY_AMT1	Quantitative. Nominal.	Amount of previous payment in September 2005 in NT dollars.	Relevant. Refers to previous payment behavior.
PAY_AMT2	Quantitative. Nominal.	Amount of previous payment in August 2005 in NT dollars.	Relevant.
PAY_AMT3	Quantitative. Nominal.	Amount of previous payment in July 2005 in NT dollars.	Relevant.
PAY_AMT4	Quantitative. Nominal.	Amount of previous payment in June 2005 in NT dollars.	Relevant.
PAY_AMT5	Quantitative. Nominal.	Amount of previous payment in May 2005 in NT dollars.	Relevant.
PAY_AMT6	Quantitative. Nominal.	Amount of previous payment in April 2005 in NT dollars.	Relevant.
default payment next month	Quantitative, Nominal.	Whether the client will default the next payment. 1 = yes 0 = no	Target feature.

Figure 3.2: A description of features from the DCCC dataset (continued)

Considering this information, it is possible to conclude that out of the 25 features (including the target feature), two need to be removed from the dataset as they are considered to be irrelevant for the purpose of the study. Please remember that features irrelevant to this study may be relevant to other studies. For instance, a study on the topic of intimate hygiene products may consider the gender feature as relevant.

Moreover, all of the features are quantitative, which means that there is no need to convert their values; all we need to do is rescale them. The target feature has also been converted into its numerical representation, where a customer who defaulted on the next payment is represented by a 1, while a customer who did not default on the payment is represented by a 0.

DATA PREPARATION

Although there are some good practices in this matter, there is not a fixed set of steps to take in order to prepare (preprocess) your dataset to develop a deep learning solution, and, most of the time, the steps to take will depend on the data at hand, the algorithm to be used, and other characteristics of the study.

> **NOTE**
>
> The process of preparing the DCCC dataset will be handled in this section, accompanied by a brief explanation. Feel free to open a Jupyter Notebook and replicate this process, considering that it will be the starting point for subsequent activities.

Nonetheless, there are some key aspects that must be dealt with before you start to train your model. Most of them you already know from the previous chapter, which will be applied to the current dataset, with the addition of the revision of class imbalance in the target feature:

- **Take a look at the data**: After reading the dataset using Pandas, print the head of the dataset. This helps to ensure that the correct dataset has been loaded. Additionally, it serves to provide evidence of transforming the dataset after preparation.

> **NOTE**
>
> To be able to read an Excel file using Pandas, make sure you have installed **xlrd** on your machine or virtual environment. To install **xlrd**, you need to run **conda install -c anaconda xlrd** on your Anaconda prompt.

The following is a snippet that's used to read the Excel file using **pandas** and print out the head of the dataset:

```
import pandas as pd

data = pd.read_excel("default of credit card clients.xls", \
                    skiprows=1)
data.head()
```

We use the **skiprows** argument to remove the first row of the Excel file, which is irrelevant as it contains a second set of headers.

On executing the code, the following result is obtained:

	ID	LIMIT_BAL	SEX	EDUCATION	MARRIAGE	AGE	PAY_0	PAY_2	PAY_3	PAY_4	...	BILL_AMT4	BILL_AMT5	BILL_AMT6	PAY_AMT1	PAY_AMT2
0	1	20000	2	2	1	24	2	2	-1	-1	...	0	0	0	0	689
1	2	120000	2	2	2	26	-1	2	0	0	...	3272	3455	3261	0	1000
2	3	90000	2	2	2	34	0	0	0	0	...	14331	14948	15549	1518	1500
3	4	50000	2	2	1	37	0	0	0	0	...	28314	28959	29547	2000	2019
4	5	50000	1	2	1	57	-1	0	-1	0	...	20940	19146	19131	2000	36681

5 rows × 25 columns

Figure 3.3: The head of the DCCC dataset

The shape of the dataset is 30,000 rows and 25 columns, which can be obtained using the following line of code:

```
print("rows:",data.shape[0]," columns:", data.shape[1])
```

- **Remove irrelevant features**: By performing analysis on each of the features, it is possible to determine that two of the features should be removed from the dataset as they are irrelevant to the purpose of the study:

```
data_clean = data.drop(columns=["ID", "SEX"])
data_clean.head()
```

The resulting dataset should contain 23 columns instead of the original 25, as shown in the following screenshot:

	LIMIT_BAL	EDUCATION	MARRIAGE	AGE	PAY_0	PAY_2	PAY_3	PAY_4	PAY_5	PAY_6	...	BILL_AMT4	BILL_AMT5	BILL_AMT6	PAY_AMT1
0	20000	2	1	24	2	2	-1	-1	-2	-2	...	0	0	0	0
1	120000	2	2	26	-1	2	0	0	0	2	...	3272	3455	3261	0
2	90000	2	2	34	0	0	0	0	0	0	...	14331	14948	15549	1518
3	50000	2	1	37	0	0	0	0	0	0	...	28314	28959	29547	2000
4	50000	2	1	57	-1	0	-1	0	0	0	...	20940	19146	19131	2000

5 rows × 23 columns

Figure 3.4: The head of the DCCC dataset after removing irrelevant features

- **Check for missing values**: Next, it is time to check whether the dataset is missing any values, and, if so, calculate the percentage of how much they represent each feature, which can be done using the following lines of code:

```
total = data_clean.isnull().sum()
percent = (data_clean.isnull().sum()\
            /data_clean.isnull().count()*100)
pd.concat([total, percent], axis=1, \
            keys=['Total', 'Percent']).transpose()
```

The first line performs a sum of missing values for each of the features of the dataset. Next, we calculate the participation of missing values for each feature. Finally, we concatenate both values calculated previously, displaying the results in a table. The results are as follows:

	LIMIT_BAL	EDUCATION	MARRIAGE	AGE	PAY_0	PAY_2	PAY_3	PAY_4	PAY_5	PAY_6	...	BILL_AMT4	BILL_AMT5	BILL_AMT6	PAY_AMT1
Total	0.0	0.0	0.0	0.0	0.0	0.0	0.0	0.0	0.0	0.0	...	0.0	0.0	0.0	0.0
Percent	0.0	0.0	0.0	0.0	0.0	0.0	0.0	0.0	0.0	0.0	...	0.0	0.0	0.0	0.0

2 rows × 23 columns

Figure 3.5: The count of missing values in the DCCC dataset

From these results, it is possible to say that the dataset is not missing any values, so no further action is required here.

- **Check for outliers**: As we mentioned in *Chapter 2*, *Building Blocks of Neural Networks*, there are several ways to check for outliers. However, in this book, we will stick to the standard deviation methodology, where values that are three standard deviations away from the mean will be considered outliers. Using the following code, it is possible to identify outliers from each feature and calculate the proportion they represent against the entire set of values:

```
outliers = {}

for i in range(data_clean.shape[1]):

    min_t = data_clean[data_clean.columns[i]].mean() \
            - (3 * data_clean[data_clean.columns[i]].std())

    max_t = data_clean[data_clean.columns[i]].mean() \
            + (3 * data_clean[data_clean.columns[i]].std())

    count = 0
    for j in data_clean[data_clean.columns[i]]:
        if j < min_t or j > max_t:
            count += 1

    percentage = count/data_clean.shape[0]

    outliers[data_clean.columns[i]] = "%.3f" % percentage

print(outliers)
```

This results in a dictionary containing each feature name as a key, with the value representing the proportion of outliers for that feature. From these results, it is possible to observe that the features containing more outliers are **BILL_AMT1** and **BILL_AMT4**, each with a participation of 2.3% out of the total instances.

This means that there are no further actions required given that the participation of outliers for each feature is too low, so they are unlikely to have an effect on the final model.

- **Check for class imbalance**: Class imbalance occurs when the class labels in the target feature are not equally represented; for instance, a dataset containing 90% of customers who did not default on the next payment, against 10% of customers who did, is considered to be imbalanced.

  There are several ways to handle class imbalance, some of which are explained here:

  Collecting more data: Although this is not always an available route, it may help balance out the classes or allow for the removal of the over-represented class without reducing the dataset severely.

  Changing performance metrics: Some metrics, such as accuracy, are not good for measuring the performance of imbalanced datasets. In turn, it is recommended to measure performance using metrics such as precision or recall for classification problems.

  Resampling the dataset: This entails modifying the dataset to balance out the classes. This can be done in two different ways: adding copies of instances within under-represented class (called oversampling) or deleting instances of the over-represented class (called undersampling).

  Class imbalance can be detected by simply counting the occurrences of each of the classes in the **target** feature, as shown here:

```
target = data_clean["default payment next month"]
yes = target[target == 1].count()
no = target[target == 0].count()

print("yes %: " + str(yes/len(target)*100) + " - no %: " \
      + str(no/len(target)*100))
```

From the preceding code, it is possible to conclude that the number of customers who defaulted on payments represents 22.12% of the dataset. These results can also be displayed in a plot using the following lines of code:

```
import matplotlib.pyplot as plt

fig, ax = plt.subplots(figsize=(10,5))
plt.bar("yes", yes)
plt.bar("no", no)
ax.set_yticks([yes,no])
plt.xticks(fontsize=15)
plt.yticks(fontsize=15)
plt.show()
```

This results in the following diagram:

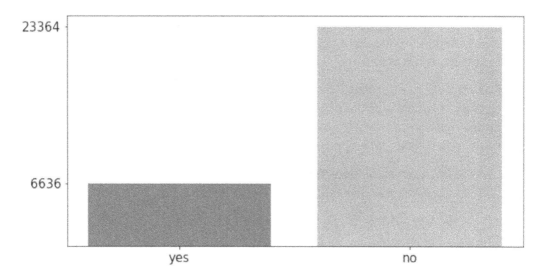

Figure 3.6: The count of classes of the target feature

In order to fix this problem, and in view of the fact that there is no more data to be added and that the performance metric is, in fact, accuracy, it is necessary to perform data resampling.

The following is a code snippet that performs oversampling over the dataset, randomly creating duplicated rows of the under-represented class:

```
data_yes = data_clean[data_clean["default payment next month"] == 1]
data_no = data_clean[data_clean["default payment next month"] == 0]

over_sampling = data_yes.sample(no, replace=True, \
                                random_state = 0)
data_resampled = pd.concat([data_no, over_sampling], \
                           axis=0)
```

First, we separate the data for each class label into independent DataFrames. Next, we use the Pandas' **sample()** function to construct a new DataFrame that contains as many duplicated instances of the under-represented class as the over-represented class's DataFrame has.

> **NOTE**
>
> Keep in mind that the first parameter of the **sample()** function **(no)** refers to the previously calculated number of items in the over-represented class.

Finally, the **concat()** function is used to concatenate the DataFrame of the over-represented class and the newly created DataFrame of the same size, in order to create the final dataset to be used in subsequent steps.

Using the newly created dataset, it is possible to, once again, calculate the participation of each class label in the target feature over the entire dataset, which should now reflect an equally represented dataset with both class labels having the same participation. The final shape of the dataset, at this point, should be equal to (46728, 23).

- **Split features from target**: We split the dataset into a features matrix and a target matrix to avoid rescaling the target values:

```
data_resampled = data_resampled.reset_index(drop=True)
X = data_resampled.drop(columns=["default payment next month"])
y = data_resampled ["default payment next month"]
```

- **Rescaling the data**: Finally, we rescale the values of the features matrix in order to avoid introducing bias to the model:

```
X = (X - X.min())/(X.max() - X.min())
X.head()
```

The result of the preceding lines of code is as follows:

	LIMIT_BAL	EDUCATION	MARRIAGE	AGE	PAY_0	PAY_2	PAY_3	PAY_4	PAY_5	PAY_6	...	BILL_AMT3	BILL_AMT4	BILL_AMT5	BILL_AMT6
0	0.080808	0.333333	0.666667	0.224138	0.2	0.2	0.2	0.2	0.2	0.2	...	0.093789	0.173637	0.095470	0.272928
1	0.040404	0.333333	0.333333	0.275862	0.2	0.2	0.2	0.2	0.2	0.2	...	0.113407	0.186809	0.109363	0.283685
2	0.040404	0.333333	0.333333	0.620690	0.1	0.2	0.1	0.2	0.2	0.2	...	0.106020	0.179863	0.099633	0.275681
3	0.040404	0.166667	0.666667	0.275862	0.2	0.2	0.2	0.2	0.2	0.2	...	0.117974	0.178407	0.100102	0.276367
4	0.494949	0.166667	0.666667	0.137931	0.2	0.2	0.2	0.2	0.2	0.2	...	0.330672	0.671310	0.559578	0.625196

5 rows × 22 columns

Figure 3.7: The features matrix after being normalized

> **NOTE**
>
> Consider that both **Marriage** and **Education** are ordinal features, meaning that they follow an order or hierarchy; when choosing a rescaling methodology, make sure to maintain order.

With the purpose of facilitating the use of the prepared dataset for the upcoming activities, both the features (**X**) and target (**y**) matrices will be concatenated into one Pandas DataFrame, which will be saved into a CSV file using the following code:

```
final_data = pd.concat([X, y], axis=1)
final_data.to_csv("dccc_prepared.csv", index=False)
```

After performing the preceding steps, the DCCC dataset has been transformed and is ready (in a new CSV file) to be used for training the model, which will be explained in the following section.

BUILDING THE MODEL

Once the problem has been defined and the data at hand has been explored and prepared, it is time to define the model. The definition of the architecture of the network, the type of layers, the loss function, and so on should be dealt with after the previous analysis. This is mainly because there is no "one-size-fits-all" approach in machine learning, and even less so in deep learning.

A regression task requires a different methodology to a classification task, as does clustering, computer vision, and machine translation. In the following section, you will find out about the key characteristics for building a model for solving a classification task, along with an explanation of how to arrive at a "good" architecture, as well as how and when to use custom modules in PyTorch.

ANNS FOR CLASSIFICATION TASKS

As seen in the *Activity 2.02, Developing a Deep Learning Solution for a Regression Problem*, from the *Chapter 2, Building Blocks of Neural Networks*, neural networks built for regression tasks use outputs as continuous values, which is why the output layer is left without an activation function and with only one output node (the real value), as in the case of a model built to predict house prices based on the characteristics of the house and the neighborhood.

In view of this, to measure performance in relation to such models, you should calculate the difference between the ground truth and the predicted value, that is, for example, calculating the distance between 125.3 (the prediction) and 126.38 (the ground truth). As we mentioned previously, there are many ways to measure this difference, with the **mean squared error** (**MSE**), or another variation, the **root mean squared error** (**RMSE**), being the most commonly used metrics.

Contrary to this, the output from a classification task is the probability of a certain set of input features belonging to each of the output labels or classes, which is done using a sigmoid (for binary classification) or a softmax (for multi-class classification) activation function. For binary classification tasks, the output layer should contain one (for sigmoid) or two (for softmax) output nodes, while for multi-class classification tasks, the output layer should be equal to the number of class labels.

This ability to calculate the likelihood of an input feature belonging to each output class, coupled with an **argmax** function, will retrieve the class with a higher probability as the final prediction.

> **NOTE**
>
> The **argmax** function, in Python, is a function capable of returning the index of the maximum value along an axis.

Considering this, the performance of the model should be a matter of whether the instances have been classified to the correct class label, rather than anything to do with measuring the distance between two values—hence the use of a different loss function (with cross-entropy being the most commonly used) for training neural networks for classification problems, as well as the use of different performance metrics, such as accuracy, precision, and recall.

A GOOD ARCHITECTURE

As has been explained throughout this chapter, it is important to understand the data problem at hand in order to determine the general topology of the neural network. Again, a regular classification problem does not require the same network architecture that's required for computer vision.

Once the data has been revised and prepared, considering that there is no right answer in terms of determining the number of hidden layers or the number of units in each layer, the best approach is to get started with an initial architecture (which can be improved to increment performance).

This is important because, with a large number of parameters to tune, sometimes, it can be difficult to commit to something and start developing the solution. However, considering that, when training neural networks, there are several ways to determine what needs to be improved once an initial architecture has been trained and tested. In fact, the whole reason for dividing your dataset into three subsets is to allow for the possibility of training the dataset with one set, measuring and fine-tuning the model with another, and, finally, measuring the performance of the final model with a final subset that has not been used before.

Considering all this, the following set of conventions and rules of thumb will be explained to aid the decision-making process for defining the initial architecture of an ANN:

- **Input layer**: This is simple enough; there is only one input layer, and its number of units is dependent on the shape of the training data. Specifically, the number of units in the input layer should be equal to the number of features that the input data contains.

- **Hidden layer**: Hidden layers can vary in quantity. ANNs can have one hidden layer, more, or none. To choose the right number, it is important to consider the following:

The simpler the data problem, the fewer hidden layers it requires. Remember that simple data problems that are linearly separable should only have one hidden layer. On the other hand, more complex data problems can, and should, be solved using many hidden layers (without limitation).

The number of hidden units should be between the number of units in the input layer and the number of units in the output layer.

- **Output layer**: Again, any ANN only has one output layer. The number of units that it contains depends on the learning task to be developed, as well as the data problem. For regression tasks, there would only be one unit, which is the predicted value. However, for classification problems, the number of units should be equal to the number of class labels available, considering that the output from the model should be the probability of a set of features belonging to each of the class labels.

- **Other parameters**: Other parameters should be left with their default values for the first configuration of the network. This is mainly because it is always good practice to test the simplest model over your data problem before considering more complex approximations that may perform equally well or worse, but will require more resources.

Once an initial architecture has been defined, it is time to train and measure the performance of the model in order to perform further analysis, which will most likely result in changes in the architecture of the network or the values of other parameters, such as changes in the learning rate or the addition of a regularization term.

PYTORCH CUSTOM MODULES

Custom modules were created by PyTorch's development team as a way to allow further flexibility to the user. Contrary to the `Sequential` container we explored in previous chapters, custom modules should be used whenever there is a desire to build more complex model architectures, or whenever you wish to have further control over the calculations that occur in each layer.

This does not mean that the custom module methodology can only be used in such scenarios. On the contrary, once you learn to use both approaches, it is a matter of preference when choosing which one (the `Sequential` container or the custom modules) to use for the less complex data problems.

Take, for instance, the following code snippet of a two-layer neural network that's been defined using the **Sequential** container:

```
import torch
import torch.nn as nn

model = nn.Sequential(nn.Linear(D_i, D_h), \
                      nn.ReLU(), \
                      nn.Linear(D_h, D_o), \
                      nn.Softmax())
```

Here, **D_i** refers to the input dimensions (the features in the input data), **D_h** refers to the hidden dimensions (the number of nodes in a hidden layer), and **D_o** refers to the output dimensions.

Using custom modules, it is possible to build an equivalent network architecture, as shown here:

```
import torch
import torch.nn as nn
import torch.nn.functional as F

class Classifier(torch.nn.Module):
    def __init__(self, D_i, D_h, D_o):
        super(Classifier, self).__init__()
        self.linear1 = torch.nn.Linear(D_i, D_h)
        self.linear2 = torch.nn.Linear(D_h, D_o)

    def forward(self, x):
        z = F.relu(self.linear1(x))
        o = F.softmax(self.linear2(z))

        return o
```

As can be seen, an input layer and an output layer are defined inside the initialization method of the class. Next, an additional method is defined where the computations are performed.

> **NOTE**
>
> For the exercises and activities in this chapter, you will need to have Python 3.7, Jupyter 6.0, Matplotlib 3.1, PyTorch 1.3, NumPy 1.17, scikit-learn 0.21, Pandas 0.25, and Flask 1.1 installed.

EXERCISE 3.01: DEFINING A MODEL'S ARCHITECTURE USING CUSTOM MODULES

Using the theory explained previously, we will define a model's architecture using the custom module's syntax:

1. Open a Jupyter Notebook and import the required libraries:

```
import torch
import torch.nn as nn
import torch.nn.functional as F
```

2. Define the necessary variables for the input, hidden, and output dimensions. Set them to **10**, **5**, and **2**, respectively:

```
D_i = 10
D_h = 5
D_o = 2
```

3. Using PyTorch's custom modules, create a class called **Classifier** and define the model's architecture so that it has two linear layers—the first one followed by a **ReLU** activation function, and the second one by a **Softmax** activation function:

```
class Classifier(torch.nn.Module):

    def __init__(self, D_i, D_h, D_o):
        super(Classifier, self).__init__()
        self.linear1 = torch.nn.Linear(D_i, D_h)
        self.linear2 = torch.nn.Linear(D_h, D_o)

    def forward(self, x):
        z = F.relu(self.linear1(x))
```

```
        o = F.softmax(self.linear2(z))

        return o
```

4. Instantiate the class and feed it with the three variables we created in *Step 2*. Print the model:

```
model = Classifier(D_i, D_h, D_o)
print(model)
```

The output from the **print** statement should be as follows:

```
Classifier(
    (linear1): Linear(in_features=10, out_features=5, bias=True)
    (linear2): Linear(in_features=5, out_features=2, bias=True)
)
```

> **NOTE**
>
> To access the source code for this specific section, please refer to https://packt.live/2VwWlgU.
>
> You can also run this example online at https://packt.live/2BrUWkD. You must execute the entire Notebook in order to get the desired result.

With that, you have successfully built a neural network architecture using PyTorch's custom modules. Now, you can move on and learn about the process of training a deep learning model.

DEFINING THE LOSS FUNCTION AND TRAINING THE MODEL

It is important to mention that the cross-entropy loss function requires the output from the network to be raw (before obtaining the probabilities through the use of the **softmax** activation function), which is why it is common to find neural network architectures for classification problems without an activation function for the output layer. Moreover, in order to get a prediction through this approach, it is necessary to apply the **softmax** activation function to the output of the network after the model has been trained.

Another way of handling this is to use the **log_softmax** activation function for the output layer instead. That way, the loss function can be defined as the negative log-likelihood loss (**nn.NLLLoss**). Finally, it is possible to get the probabilities for a set of features belonging to each class label by taking the exponential from the output of the network. This is the approach that will be used in the activities in this chapter.

Once the model architecture has been defined, the next step would be to code the section in charge of training the model over the training data, as well as measuring its performance over both the training and validation sets.

The code following these step-by-step instructions that we have discussed is as follows:

```
model = Classifier()
criterion = nn.NLLLoss()
optimizer = optim.Adam(model.parameters(), lr=0.005)

epochs = 10
batch_size = 100
```

As can be seen in the preceding snippet, the first step is to define all the variables that will be used while training the network.

Next, a first **for** loop is used to go through the number of epochs we defined previously.

Keep in mind that **epochs** refer to the number of times that the entire dataset is passed forward and backward through the network architecture. **batch_size** refers to the number of training examples in a single batch (a slice of the dataset). Finally, **iterations** refer to the number of batches required to complete one epoch.

A second **for** loop is used to go through each batch of the total dataset until an epoch is completed. Inside this loop, the following computations occur:

1. The model is trained over a batch of the training set. A prediction is obtained here.

2. The loss is calculated by comparing the prediction from the previous step and the labels from the training set (ground truth).

3. The gradients are set to zero and calculated again for the current step.

4. The parameters of the network are updated based on the gradients.

5. The accuracy of the model over the training data is calculated as follows:

   Get the exponential of the predictions of the model in order to obtain the probabilities of a given piece of data belonging to each class label.

   Use the **topk()** method to get the class label with a higher probability.

   Using scikit-learn's metric section, calculate the accuracy, precision, or recall. You can also explore other performance metrics.

Once all batches of the training data have been fed to the model, the calculation of gradients is turned off in order to verify the performance of the current model over the validation data, which occurs as follows:

1. The model performs a prediction for the data in the validation set.

2. The loss function is calculated by comparing the previous prediction with the labels from the validation set.

3. The accuracy is calculated over the validation set. To calculate the accuracy of the model over the validation set, use the same set of steps that you did for the same calculation over the training data:

> **NOTE**
>
> The following code snippet won't run by itself. You will need to have a dataset loaded and divided into the different sets and a network architecture defined and instantiated. The loss function and optimization algorithm (which were explained in the previous sections of this chapter) also need to be defined.

```
train_losses, dev_losses, \
train_acc, dev_acc= [], [], [], []

for e in range(epochs):
    X, y = shuffle(X_train, y_train)
    running_loss = 0
    running_acc = 0
```

```
iterations = 0

for i in range(0, len(X), batch_size):
    iterations += 1
    b = i + batch_size
    X_batch = torch.tensor(X.iloc[i:b,:].\
                           values).float()
    y_batch = torch.tensor(y.iloc[i:b].values)

    pred = model(X_batch)
    loss = criterion(pred, y_batch)
    optimizer.zero_grad()
    loss.backward()
    optimizer.step()

    running_loss += loss.item()
    ps = torch.exp(pred)
    top_p, top_class = ps.topk(1, dim=1)
    running_acc += accuracy_score(y_batch, top_class)

dev_loss = 0
acc = 0

with torch.no_grad():
    pred_dev = model(X_dev_torch)
    dev_loss = criterion(pred_dev, y_dev_torch)
    ps_dev = torch.exp(pred_dev)
    top_p, top_class_dev = ps_dev.topk(1, dim=1)
    acc = accuracy_score(y_dev_torch, top_class_dev)

train_losses.append(running_loss/iterations)
dev_losses.append(dev_loss)
train_acc.append(running_acc/iterations)
dev_acc.append(acc)

print("Epoch: {}/{}.. ".format(e+1, epochs), \
      "Training Loss: {:.3f}.. "\
```

```
                .format(running_loss/iterations),\
        "Validation Loss: {:.3f}.. "\
        .format(dev_loss),\
        "Training Accuracy: {:.3f}.. "\
        .format(running_acc/iterations),\
        "Validation Accuracy: {:.3f}".format(acc))
```

The preceding code snippet will print the loss and accuracy for both sets of data. In the following activity, all of the concepts we explained regarding building and training a DNN will be put into practice.

ACTIVITY 3.01: BUILDING AN ANN

For this activity, using the previously prepared dataset, we will build a four-layer model that is able to determine whether a client will default the next payment. To do this, we will use the custom module's methodology.

Let's look at the following scenario: you work at a data science boutique that specializes in providing machine/deep learning solutions to banks all around the world. They have recently taken on a project for a bank that wishes to be able to predict the payments that will or will not be received the following month. The exploratory data analysis team has already prepared the dataset for you and they have asked you to build the model and calculate the accuracy of the model. Follow these steps to complete this activity:

1. Import the following libraries:

```
import pandas as pd
import numpy as np
from sklearn.model_selection import train_test_split
from sklearn.utils import shuffle
from sklearn.metrics import accuracy_score
import torch
from torch import nn, optim
import torch.nn.functional as F
import matplotlib.pyplot as plt
```

> **NOTE**
>
> Even with the use of a seed, the exact results of this activity will not be reproducible, considering that the training sets are shuffled before each epoch.

2. Read the previously prepared dataset, which should have been named **dccc_prepared.csv**.

3. Separate the features from the target.

4. Using scikit-learn's **train_test_split** function, split the dataset into training, validation, and testing sets. Use a 60:20:20 split ratio. Set **random_state** to **0**.

5. Convert the validation and testing sets into tensors, considering that the features' matrix should be of the float type, while the target matrix should not. Leave the training sets unconverted for the moment as they will undergo further transformations.

6. Build a custom module class for defining the layers of the network. Include a forward function that specifies the activation functions that will be applied to the output of each layer. Use **ReLU** for all layers except for the output, where you should use **log_softmax**.

7. Instantiate the model and define all the variables required to train the model. Set the number of epochs to **50** and the batch size to **128**. Use a learning rate of **0.001**.

8. Train the network using the training set's data. Use the validation sets to measure performance. To do this, save the loss and the accuracy for both the training and validation sets in each epoch.

> **NOTE**
>
> The training process may take several minutes, depending on your resources. Adding print statements is good practice if you wish to see the progress of the training process.

9. Plot the loss of both sets.

10. Plot the accuracy of both sets.

The final plot will look as follows:

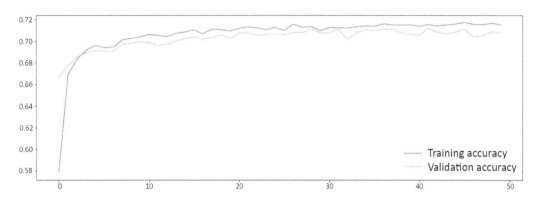

Figure 3.8: A plot displaying the accuracy of the sets

> **NOTE**
>
> The solution to this activity can be found on page 245.

You have now successfully programmed and trained a four-layer neural network capable of performing inferences based on historical data. Next, you will learn how to improve a model's performance in order to produce trustworthy inferences over unseen data.

DEALING WITH AN UNDERFITTED OR OVERFITTED MODEL

Building a deep learning solution is not just a matter of defining an architecture and then training a model using the input data; on the contrary, most would agree that this is the easy part. The art of creating high-tech models entails achieving high levels of accuracy that surpass human performance. This section will introduce the topic of error analysis, which is commonly used to diagnose a trained model to discover what actions are more likely to have a positive impact on the performance of the model.

ERROR ANALYSIS

Error analysis refers to the initial analysis of the error rate over the training and validation sets of data. This analysis is then used to determine the best course of action to improve the performance of the model.

In order to perform error analysis, it is necessary to determine the Bayes error (also known as the irreducible error), which is the minimum achievable error. Several decades ago, the Bayes error was equivalent to human error, meaning that the conceived minimum level of error was the one that experts could achieve.

Nowadays, with the improvements in technology and algorithms, this value has become increasingly difficult to estimate since machines are capable of surpassing human performance; there is no measure of how much better they can do in comparison to humans as we can only understand as far as our capacity goes.

It is common for the Bayes error to be set equal to the human error, initially, in order to perform error analysis. However, this limitation is not immutable, and researchers hold that surpassing human performance should be an end goal as well.

The process of performing error analysis is as follows:

1. Calculate the metric of choice to measure the performance of the model. This measure should be calculated over both the training and validation sets of data.

2. Using this measure, calculate the error rate of each of the sets by subtracting the performance metric that was previously calculated from 1. Take, for instance, the following equation:

$$error_{training} = 1 - accuracy_{training}$$

Figure 3.9: The equation to calculate the error rate of the model over the training set

3. Subtract the Bayes error from the training set error, **A)** . Save the difference, which will be used for further analysis.

4. Subtract the training set error from the validation set error, **B)** , and save the value of the difference.

5. Take the differences calculated in Steps 3 and 4 and use the following set of rules:

   If the difference calculated in *Step 3* is higher than the one calculated in *Step 4*, the model is underfitted, also known as suffering from high bias.

If the difference calculated in *Step 4* is higher than the one calculated in *Step 3*, the model is overfitted, also known as suffering from high variance, as shown in the following diagram:

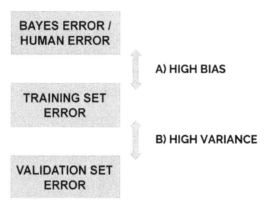

Figure 3.10: Diagram showing how to perform error analysis

These rules do not indicate that the model could only be suffering from one of the issues just mentioned (high bias or high variance), but rather that the one detected through error analysis is having a greater effect on the performance of the model, which means that fixing it will improve the performance to a greater degree.

Let's explain how to deal with each of these issues:

- **High bias**: An underfitted model, or a model suffering from high bias, is a model that is not capable of understanding the training data, and, hence, it is not able to uncover patterns and generalize with other sets of data. This means that the model does not perform well over any set of data.

  To decrease the high bias affecting the model, it is recommended that you define a bigger/deeper network (more hidden layers) or train for more iterations. By adding more layers and increasing the training time, the network has more resources to discover the patterns that describe the training data.

- **High variance**: An overfitted model, or a model suffering from high variance, is a model that is having trouble generalizing the training data; it is learning the details of the training data too well (this means that through the training process, the model learned the information from the training set too well, which means that it is unable to generalize to other sets of data), including its outliers. This means that the model is performing too well over the training data, but poorly over other sets of data.

This is typically handled by adding more data to the training set, or by adding a regularization term to the loss function. The first approach aims to force the network to generalize to the data rather than understand the details of a small number of examples. The second approach, on the other hand, penalizes the inputs that have higher weights in order to overlook outlier values and consider all values equally.

In view of this, dealing with one condition that is affecting the model may cause another one to appear or increase. For instance, a model suffering from high bias, after being treated, may improve its performance over the training data but not over the validation data, which means that the model will have started to suffer from high variance and will require another set of remedial actions to be taken.

Once the model has been diagnosed and the necessary measures have been taken to improve performance, the best models should be selected for a final test. Each of these models should be used to perform predictions over the testing set (the only set that does not have an effect on building the model).

Considering this, it is possible to select the final model as the one that performs best over the testing data. This is mainly because the performance over the testing data serves as an indicator of the model's performance on future unseen sets of data, which is the ultimate goal.

EXERCISE 3.02: PERFORMING ERROR ANALYSIS

Using the accuracy metric we calculated in the previous activity, in this activity, we will perform error analysis, which will help us determine the actions to be performed over the model in the upcoming activity. Follow these steps to complete this exercise:

> **NOTE**
>
> This exercise does not require any coding to be done, but rather an analysis of the previous activity's results.

1. Assuming a Bayes error of **0.15**, perform error analysis and diagnose the model:

```
Bayes error (BE) = 0.15
Training set error (TSE) = 1 - 0.716 = 0.284
Validation set error (VSE) = 1 - 0.71 = 0.29
```

The values that are being used as the accuracy of both sets (**0.716** and **0.71**) are the ones that were obtained during the last iteration of *Activity 3.01, Building an ANN*:

```
High bias = TSE - BE = 0.134
High variance = VSE - TSE = 0.014
```

According to this, the model is suffering from high bias, meaning that the model is underfitted.

2. Determine the actions necessary to improve the accuracy of the model.

    To improve the model's performance, the two courses of action that can be followed are incrementing the number of epochs and increasing the number of hidden layers and/or the number of units (neurons in each layer).

    In accordance with this, a set of tests can be performed in order to arrive at the best result.

With that, you have successfully performed error analysis. This methodology is key to developing state-of-the-art deep learning solutions that perform outstandingly well over unseen data.

ACTIVITY 3.02: IMPROVING A MODEL'S PERFORMANCE

For this activity, we will implement the actions we defined in the exercise to reduce the high bias that is affecting the performance of the model. Consider the following scenario: your teammates are impressed with the work you have delivered and the way that your code is organized, but they have asked you to try to improve the performance to 80%, considering that this is what they promised to the client. Follow these steps to complete this activity:

> ### NOTE
>
> Use a different Jupyter Notebook for this activity. There, you will load the dataset again and perform similar steps as in the previous activity, with the difference being that the training process will be done several times to train different architectures and training times.

1. Import the same libraries as in the previous activity.

2. Load the data and split the features from the target. Next, split the data into three subsets (training, validation, and testing) using a 60:20:20 split ratio. Finally, convert the validation and testing sets into PyTorch tensors, just as you did in the previous activity.

3. Considering that the model is suffering from a high bias, the focus should be on increasing the number of epochs or increasing the size of the network by adding additional layers or units to each layer. The aim should be to approximate the accuracy over the testing set to 80%.

> **NOTE**
>
> There is no correct way to choose which test to carry out first, so be creative and analytical. If changes in your model's architecture reduce or eliminate the high bias but introduce high variance, then consider, for instance, keeping the changes but adding a measure to combat the high variance.

4. Plot the loss and accuracy for both sets of data.

5. Using the best-performing model, perform a prediction over the testing set (which should not have been used during the fine-tuning process). Compare the prediction to the ground truth by calculating the accuracy of the model over this set.

Expected output: The accuracy obtained through the model architecture and the parameters that are defined here should be around 80%.

> **NOTE**
>
> The solution to this activity can be found on page 250.

With that, you have successfully improved your model's performance using error analysis. Next, you will learn how to deploy your model in order to make use of it in production environments.

DEPLOYING YOUR MODEL

By now, you have learned and put into practice the key concepts and tips for building exceptional deep learning models for regular regression and classification problems. In real life, models are not just built for learning purposes. On the contrary, when training models for purposes other than research, the main idea is to be able to reuse them in the future to perform predictions over new data that, although the model was not trained on, the model should perform similarly well with.

In a small organization, the ability to serialize and deserialize models suffices. However, when models are to be used by large corporations, by users, or to alter a massively important and large task, it is a better practice to convert the model into a format that can be used in most production environments (such as APIs, websites, and online and offline applications).

In this section, we will learn how to save and load models, as well as how to use PyTorch's most recent feature to convert our model into a C++ application that is highly versatile. We will also learn how to create an API to make use of a trained model.

SAVING AND LOADING YOUR MODEL

As you might imagine, retraining a model every time it is to be used is highly impractical, especially considering that most deep learning models may take quite some time to train (depending on your resources).

Instead, models in PyTorch can be trained, saved, and reloaded to either perform further training or to make inferences. This can be achieved considering that the parameters (weights and biases) for each layer in the PyTorch models are saved into the **state_dict** dictionary.

Here, a step-by-step guide is provided on how to save and load a trained model:

1. Originally, a checkpoint of a model will only include the model's parameters. However, when loading the model, this is not the only information that's required. Depending on the arguments that your classifier takes in (that is, the class containing the network architecture), it may be necessary to save further information, such as the number of **input** units. Considering this, the first step is to define the information to be saved:

```
checkpoint = {"input": X_train.shape[1], \
              "state_dict": model.state_dict()}
```

This will save the number of units in the input layer into the checkpoint, which will come in handy when loading the model.

2. Save the model using PyTorch's **save()** function:

```
torch.save(checkpoint, "checkpoint.pth")
```

The first argument refers to the dictionary we created previously, while the second argument is the filename to be used.

3. Using a text editor of your choice, create a Python file that imports PyTorch libraries and contains the class that creates the network architecture of your model. This is done so that you can conveniently load the model into a new worksheet, separate from the one you used to train the model.

4. To load the model, let's create a function that will perform three main actions:

```
def load_model_checkpoint(path):
    checkpoint = torch.load(path)

    model = final_model.Classifier(checkpoint["input"], \
                                   checkpoint["output"], \
                                   checkpoint["hidden"])

    model.load_state_dict(checkpoint["state_dict"])
    return model

model = load_model_checkpoint("checkpoint.pth")
```

This function receives the path to the saved model file (the checkpoint) as input. First, the checkpoint is loaded. Next, a model is instantiated using the network's architecture that was saved into the Python file. Here, **final_model** refers to the name of the Python file, which should have been imported into the new worksheet, while **Classifier()** refers to the name of the class saved in that file. This model will have randomly initialized parameters. Finally, the parameters from the checkpoint are loaded into the model.

When called, this function returns the trained model, which can now be used for further training or to perform inferences.

PYTORCH FOR PRODUCTION IN C++

As per the name of the framework, PyTorch's primary interface is the Python programming language. This is mainly due to the preference for this programming language on the part of many users, thanks to the language's dynamism and ease of use when it comes to developing machine learning solutions.

Nevertheless, in some scenarios, Python properties become unfavorable. This is precisely the case for models developed for production, where other programming languages have proven to be more useful. Such is the case with C++, which is widely used for production purposes with machine/deep learning solutions.

Given this, PyTorch recently proposed an easy approach to allow users to enjoy the benefits of both worlds. While they get to continue programming in a Pythonic nature, it is now possible to serialize your model into a representation that can be loaded and executed from C++, with no dependency on Python. This representation is called TorchScript.

Converting a PyTorch model into a TorchScript is done through PyTorch's **Just-In-Time (JIT)** compiler module. This is achieved by passing your model, along with an example input, through the `torch.jit.trace()` function, as shown here:

```
traced_script = torch.jit.trace(model, example)
```

Keep in mind that the variable named **model** should contain the previously trained model, while the variable named **example** should contain the set of features that you wish to feed to your model in order to perform a prediction. This will return a script module, which can be used as a regular PyTorch module, as shown here:

```
prediction = traced_script(input)
```

The preceding code will return the output that was obtained from running the input data through your model.

BUILDING AN API

An **Application Programming Interface** (**API**) consists of a program specifically created to be used by other programs (as opposed to a website or an interface, which are created to be manipulated by humans). According to this, APIs are used when creating deep learning solutions to be used in production environments as they allow for the information derived from running the model (a prediction, for instance) to be accessed through other means (a website, for instance).

In this section, we will explore the creation of a web API (an API that shares information with other programs through the internet). The functionality of this API will be to load the previously saved model and make a prediction from a given set of features. This prediction can be accessed by the program that makes an HTTP request to the API.

The key terminology is explained as follows:

- **Hypertext Transfer Protocol** (**HTTP**): This is the primary means of transferring data on the web. It works using **methods**, which help to determine the way in which data is transferred. The two most commonly used methods are explained as follows:

  POST: This method allows you to send data from the client (a web browser or the platform that is making the request to the API) to the server (the program to be run using that information) in the message body.

  GET: Contrary to the **POST** method, this method sends data as part of the URL, which may be inconvenient when sending large quantities of data.

- **Flask**: This is a library that's been developed for Python that allows you to create APIs (among other things).

To create a simple web API using Flask, perform the following steps:

1. Import the necessary libraries:

```
import flask
from flask import request
import torch
import final_model
```

2. Initialize the Flask app:

```
app = flask.Flask(__name__)
app.config["DEBUG"] = True
```

    The **DEBUG** configuration is set to **True** during development but should be set to **False** when in production.

3. Load a previously trained model:

```
def load_model_checkpoint(path):
    checkpoint = torch.load(path)
    model = final_model.Classifier(checkpoint["input"])
    model.load_state_dict(checkpoint["state_dict"])
    return model


model = load_model_checkpoint("checkpoint.pth")
```

4. Define the route where the API will be accessible, as well as the method(s) that can be used for sending information to the API in order to perform an action. This syntax is called a decorator and should be located immediately before a function:

```
@app.route('/prediction', methods=['POST'])
```

5. Define a function that performs the desired action. In this case, the function will take the information that was sent to the API and feed it to the previously loaded model to perform a prediction. Once the prediction has been obtained, the function should return a response, which will be displayed as the result of the API request:

```
def prediction():

    body = request.get_json()
    example = torch.tensor(body['data']).float()
    pred = model(example)
    pred = torch.exp(pred)
    _, top_class_test = pred.topk(1, dim=1)
    top_class_test = top_class_test.numpy()

    return {"status":"ok", "result":int(top_class_test[0][0])}
```

6. Run the Flask app. The following command makes it possible to start making use of the web API:

```
app.run(debug=True, use_reloader=False)
```

Again, **debug** is set to **True** during development. The **use_reloader** parameter is set to **False** to allow the app to run over a Jupyter Notebook. However, it is not recommended to run the application from a Jupyter Notebook; this is only being done for teaching purposes. In real life, **use_reloader** should be set to **True** and the application should be run as a Python file through a Command Prompt instance or Terminal.

EXERCISE 3.03: CREATING A WEB API

Using Flask, we will create a web API that will receive some data when it's called and will return a piece of text that will be displayed in a browser. Follow these steps to complete this exercise:

1. In a Jupyter Notebook, import the required libraries:

```
import flask
from flask import request
```

2. Initialize the Flask app:

```
app = flask.Flask(__name__)
app.config["DEBUG"] = True
```

3. Define the route of your API so that it's **/<name>**. Set the method to **GET**. Next, define a function that takes in an argument (**name**) and returns a string that contains an **h1** tag with the word **HELLO**, followed by the argument received by the function:

```
@app.route('/<name>', methods=['GET'])
def hello(name):
    return "<h1>HELLO {}</h1>".format(name.upper())
```

4. Run the Flask app:

```
app.run(debug=True, use_reloader=False)
```

The output from the preceding line of code will look as follows:

```
* Serving Flask app "__main__" (lazy loading)
* Environment: production
  WARNING: This is a development server. Do not use it in a production deployment.
  Use a production WSGI server instead.
* Debug mode: on
* Running on http://127.0.0.1:5000/ (Press CTRL+C to quit)
127.0.0.1 - - [09/Dec/2019 18:45:30] "GET / HTTP/1.1" 404 -
127.0.0.1 - - [09/Dec/2019 18:45:31] "GET /favicon.ico HTTP/1.1" 200 -
127.0.0.1 - - [09/Dec/2019 18:45:36] "GET /your_name HTTP/1.1" 200 -
127.0.0.1 - - [09/Dec/2019 18:45:36] "GET /favicon.ico HTTP/1.1" 200 -
```

Figure 3.11: Warnings after executing the code

It contains several warnings that specify the conditions of the Flask app that is running, as well as a line of text containing a URL similar to this:

```
http://127.0.0.1:5000/
```

5. Copy the URL into a browser, followed by your name, as follows:

```
http://127.0.0.1:5000/your_name
```

Press *Enter* and a simple website should load, which should look similar to the following one:

Figure 3.12: Result from performing a request to the API

> **NOTE**
>
> To access the source code for this specific section, please refer to https://packt.live/3icVgEF.
>
> This section does not currently have an online interactive example, and will need to be run locally.

With that, you have successfully created a web API. This ability will be key to you being able to make use of your model in production environments by allowing easy communication between your model and the user. In the next activity, you will put the different concepts of deploying a model you've learned about in this chapter into practice.

ACTIVITY 3.03: MAKING USE OF YOUR MODEL

For this activity, save the model you created in the previous activity. Moreover, the saved model will be loaded into a new notebook for use. Next, we will convert the model into a serialized representation that can be executed on C++, as well as create a Flask API. Let's look at the following scenario: everyone is very happy with your commitment to improving the model, as well as the final version of it, so they have asked you to save the model and convert it into a format that they can use to build an online application for the client. Follow these steps to complete this activity:

> **NOTE**
>
> This activity will make use of three Jupyter Notebooks. First, we will use the same notebook from the previous activity to save the final model. Next, we will open a new notebook, which will be used to load the saved model. Finally, a third notebook will be used to create an API.

1. Open the Jupyter Notebook that you used for *Activity 3.02, Improving a Model's Performance*.

2. Copy the class containing the architecture of your best-performing model and save it in a Python file. Make sure to import PyTorch's required libraries and modules. Name it **final_model.py**.

3. In the Jupyter Notebook, save the best-performing model. Make sure to save the information pertaining to the input units, along with the parameters of the model. Name it **checkpoint.pth**.

4. Open a new Jupyter Notebook.

5. Import PyTorch, as well as the Python file we created in *Step 2*.

6. Create a function that loads the model.

7. Perform a prediction by inputting the following tensor into your model:

```
torch.tensor([[0.0606, 0.5000, 0.3333, 0.4828, 0.4000, \
               0.4000, 0.4000, 0.4000, 0.4000, 0.4000, \
               0.1651, 0.0869, 0.0980, 0.1825, 0.1054, \
               0.2807, 0.0016, 0.0000, 0.0033, 0.0027, \
               0.0031, 0.0021]]).float()
```

8. Convert the model using the JIT module.

9. Perform a prediction by inputting the tensor from *Step 7* into the traced script of your model.

10. Open a new Jupyter Notebook and import the libraries required to create an API using Flask, as well as the libraries to load the saved model.

11. Initialize the Flask app.

12. Define a function that loads the saved model and instantiates the model.

13. Define the route of the API so that it's **/prediction** and the method so that it's **POST**. Then, define the function that will receive the **POST** data and feed it to the model to perform a prediction.

14. Run the Flask app.

The application will look as follows after running it:

Figure 3.13: A screenshot of the app after running it

> **NOTE**
>
> The solution to this activity can be found on page 257.

SUMMARY

After covering most of the theoretical knowledge in the previous chapters, this chapter used a real-life case study to cement our knowledge. The idea is to encourage learning through practice with a hands-on approach.

The chapter started off by explaining the influence of deep learning on a wide range of industries where accuracy is required. One of the main industries driving deep learning's growth is banking and finance, where such algorithms are being used in domains such as the evaluation of loan applications, the detection of fraud, and the evaluation of past decision-making to predict future behavior, mainly due to the algorithm's ability to supersede human performance in these respects.

This chapter used a real-life dataset from a Taiwanese bank, with the objective of predicting whether a client would default on a payment. This chapter started developing a solution to this by explaining the importance of defining the what, why, and how of any data problem, as well as analyzing the data at hand to make the best use of it.

Once the data was prepared according to the problem definition, we explored the idea of defining a "good" architecture. Even though there are a couple of rules of thumb that can be considered, the main takeaway was to build an initial architecture without overthinking it, in order to get some results that can be used to perform error analysis to improve the model's performance.

The idea of error analysis entails analyzing the error rate of the model over the training and validation sets in order to determine whether the model is suffering more from either high bias or high variance. This diagnosis of the model is then used to alter the model's architecture and some of the learning parameters, which will result in an improvement of performance.

Finally, we explored three main approaches to making use of the best-performing model. The first approach consists of saving the model and then reloading it into any coding platform so that we can continue training or perform inferences. The second approach is mainly used to launch the model into production and is achieved by making use of PyTorch's JIT module, which creates a serialized representation of the model that can be run on C++. Finally, the third approach consists of the creation of an API that can be accessed by other programs so that it can send and receive information to/from the model.

In the next chapter, we'll focus on solving an image classification task using CNNs.

4

CONVOLUTIONAL NEURAL NETWORKS

OVERVIEW

This chapter explains the process of training **convolutional neural networks (CNNs)**—that is, the computations that occur in the different layers that can be typically found in a CNN architecture and their purpose in the training process. You will learn how to improve a computer vision model's performance by applying data augmentation and batch normalization to a model. By the end of this chapter, you will be able to use CNNs to solve image classification problems using PyTorch. This will be the starting point for implementing other solutions in the domain of computer vision.

INTRODUCTION

In the previous chapter, the most traditional neural network architecture was explained and applied to a real-life data problem. In this chapter, we will explore the different concepts of CNNs, which are mainly used to solve computer vision problems (that is, image processing).

Even though all neural network domains are popular nowadays, CNNs are probably the most popular of all neural network architectures. This is mainly because, although they work in many domains, they are particularly good at dealing with images, and advances in technology have allowed the collection and storage of large amounts of images, which makes it possible to tackle a great variety of today's challenges using images as input data.

From image classification to object detection, CNNs are being used to diagnose cancer patients and detect fraud in systems, as well as to construct well-thought-out self-driving vehicles that will revolutionize the future.

This chapter will focus on explaining the reasons why CNNs outperform other architectures when dealing with images, as well as explaining the building blocks of their architecture in greater detail. It will cover the main coding structure for building a CNN to solve an image classification data problem.

Moreover, we will explore the concepts of data augmentation and batch normalization, which will be used to improve the performance of the model. The ultimate goal of this chapter is to compare the results of three different approaches in order to tackle an image classification problem using CNNs.

> **NOTE**
>
> All the code present in this chapter can be found at:
> https://packt.live/3bg0KKP.

BUILDING A CNN

CNNs are the ideal architecture when dealing with an image data problem. However, they are often underused as they are typically used for image classification tasks, even though their abilities extend to other domains in the field of image processing. This chapter will not only explain the reasons why CNNs are so good at understanding images but will also identify the different tasks that can be tackled, as well as provide some examples of real-life applications.

Moreover, this chapter will explore the different building blocks of CNNs and their application using PyTorch to ultimately build a model that solves a data problem using one of PyTorch's datasets for image classification.

WHY ARE CNNS USED FOR IMAGE PROCESSING?

An image is a matrix of pixels, so why not just flatten the matrix into a vector and process it using a traditional neural network architecture? The answer is that, even with the simplest image, there are some pixel dependencies that alter the meaning of the image. For instance, the representation of a cat's eye, a car tire, or even the edge of an object is constructed out of several pixels laid out in a certain way. If we flatten the image, these dependencies are lost, and so is the accuracy of a traditional model:

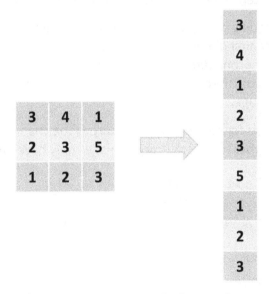

Figure 4.1: Representation of a flattened matrix

A CNN is capable of capturing the spatial dependencies of images since it processes them as matrices and analyzes entire chunks of an image at a time, depending on the size of the filter. For example, a convolutional layer using a filter of size 3 x 3 will analyze 9 pixels at a time until it has covered the entire image.

Each chunk of the image is given a set of parameters (weight and bias) that will refer to the relevance of that set of pixels to the entire image, depending on the filter at hand. This means that a vertical edge filter will assign greater weights to the chunks of the image that contain a vertical edge. According to this, by reducing the number of parameters and by analyzing the image in chunks, CNNs are capable of rendering a better representation of the image.

THE IMAGE AS INPUT

As we mentioned previously, the typical inputs of a CNN are images in the form of matrices. Each value of the matrix represents a pixel in the image, where the number is determined by the intensity of the color, with values ranging from 0 to 255.

In grayscale images, white pixels are represented by the number 255 and black pixels by the number 0. Gray pixels are any number in between, depending on the intensity of the color; the lighter the gray, the closer the number is to 255.

Colored images are usually represented using the RGB system, which represents each color as the combination of red, green, and blue. Here, each pixel will have three dimensions, one for each color. The values in each dimension will range from 0 to 255. Here, the more intense the color, the closer the number is to 255.

According to the preceding paragraph, the matrix of a given image is three-dimensional. Here, the first dimension refers to the height of the image (in the number of pixels), the second dimension refers to the width of the image (in the number of pixels), and the third dimension is known as the channel and refers to the color scheme of the image.

The number of channels for colored images is three (one channel for each color in the RGB system). On the other hand, gray-scaled images only have one channel:

Figure 4.2: Matrix representation of an image – to the left,
a colored image; to the right, a gray-scaled image

In contrast to text data, images that are fed into CNNs do not require much preprocessing. Images are usually fed as they are, with the most common changes being as follows:

- Normalizing the pixel values in order to speed up the learning process and improve performance

- Downsizing images (that is, reducing their width and length) to speed up the learning process

The simplest way to normalize inputs is to take the value of each pixel and divide it by 255 so that we end up with values ranging between 0 and 1. Nevertheless, different methodologies are used to normalize an image, such as the mean-centering technique. The decision to choose one or the other is, most of the time, a matter of preference; however, when using pretrained models, it is highly recommended that you use the same technique that you used to train the model the first time. This information is often available in the documentation of the pretrained model.

APPLICATIONS OF CNNS

Although CNNs are mainly used for computer vision problems, it is important to mention their ability to solve other learning problems, mainly with regard to analyzing sequences of data. For instance, CNNs have been known to perform well on sequences of text, audio, and video, sometimes in combination with other network architectures, or by converting the sequences into images that can be processed by CNNs. Some of the specific data problems that can be tackled using CNNs with sequences of data are machine translations of text, natural language processing, and video frame tagging, among many others.

There are different tasks that CNNs can perform that apply to all supervised learning problems; however, this chapter will focus on computer vision. The following is a brief explanation of each of these tasks, along with a real-life example of each of them.

CLASSIFICATION

This is the most commonly known task in computer vision. The main idea is to classify the general contents of an image into a set of categories, known as labels.

For instance, classification can determine whether an image is of a dog, a cat, or any other animal. This classification is done by outputting the probability of the image belonging to each of the classes, as seen in the following image:

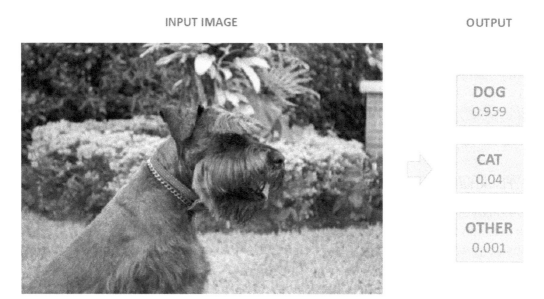

Figure 4.3: Classification task

LOCALIZATION

The main purpose of localization is to generate a bounding box that describes the object's location in the image. The output consists of a class label and a bounding box.

This task can be used in sensors to determine whether an object is to the left or right of the screen:

Figure 4.4: Localization task

DETECTION

This task consists of performing object localization on all the objects in the image. The output consists of multiple bounding boxes, as well as multiple class labels (one for each box).

This task is used in the construction of self-driving cars, with the objective of being able to locate traffic signs, the road, other cars, pedestrians, and any other object that may be relevant to ensure a safe driving experience:

Figure 4.5: Detection task

SEGMENTATION

The task here is to output both a class label and an outline of each object present in the image. This is mainly used to mark important objects of an image for further analysis.

For instance, this task can be used to strictly delimit the area corresponding to a tumor in an image of the lung of a patient. The following figure depicts how the object of interest is outlined and assigned a label:

Figure 4.6: Segmentation task

From this section onward, this chapter will focus on training a model to perform image classification using one of PyTorch's image datasets.

THE BUILDING BLOCKS OF CNNS

A deep convolutional network is one that takes an image as input and passes it through a series of **convolutional layers** with filters, **pooling layers**, and **fully connected (FC)** layers to finally apply a `softmax` activation function that classifies the image into a class label. This form of classification, as with ANNs, is performed by calculating the probability of the image belonging to each of the class labels by giving each class label a value between zero and one. The class label with the higher probability is the one that's selected as the final prediction for that image.

The following is a detailed explanation of each of these layers, along with code examples of how to define such layers in PyTorch.

CONVOLUTIONAL LAYERS

This is the first step in extracting features from an image. The objective is to maintain the relationship between nearby pixels by learning the features over small sections of the image.

A mathematical operation occurs in this layer, where two inputs are given (the image and the filter) and an output is obtained. As we explained previously, this operation consists of convolving the filter and a section of the image of the same size as the filter. This operation is repeated for all subsections of the image.

> **NOTE**
>
> Revisit the *Introduction to CNNs* section of *Chapter 2, Building Blocks of Neural Networks*, for a reminder of the exact calculation that's performed between the input and the filter.

The resulting matrix will have a shape that depends on the shapes of the inputs, where an image matrix of size (h x w x c) and a filter of size (f_h x f_w x c) will output a matrix according to the following equation:

$$output\ height = h - f_h + 1$$

$$output\ width = w - f_w + 1$$

$$output\ depth = 1$$

Figure 4.7: Output height, width, and depth of a convolutional layer

Here, h refers to the height of the input image, w is the width, c refers to the depth (also known as channels), and f_h and f_w are values that are set by the user concerning the size of the filter.

The following diagram depicts this dimension conversion in the form of matrices, where the matrices to the left represent a colored image, the matrices in the center represent a single filter that is being applied to all the channels of the image, and the matrix to the right is the output from the computation of the image and the filter:

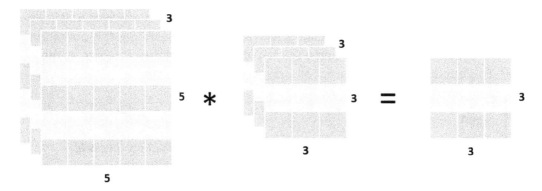

Figure 4.8: Dimensions of the input, filter, and output

It is important to mention that in a single convolutional layer, several filters can be applied to the same image, all of the same shape. Considering this, the output shape of a convolutional layer that applies two filters to its input, in terms of its depth, is equal to two, as seen in the following diagram:

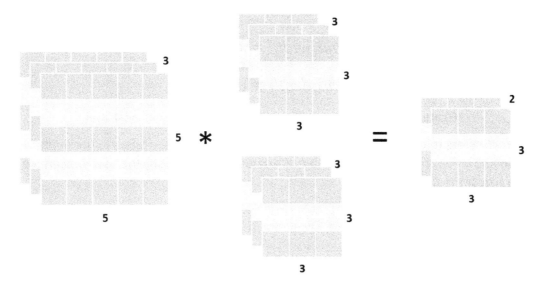

Figure 4.9: Convolutional layer with two filters

Each of these filters will perform a different operation in order to discover the different features of an image. For instance, in a single convolutional layer with two filters, these operations could be vertical edge detection and horizontal edge detection. As the network grows in terms of the number of layers, the filters will perform more complex operations that make use of previously detected features—for example, the detection of the outline of a person by using the inputs from the edge detectors.

Filters typically increase in each layer. This means that, while the first convolutional layer has 8 filters, it is common to create the second convolutional layer so that it has twice that number (16), the third so that it has twice that number again (32), and so on.

However, it is important to mention that, in PyTorch, as in many other frameworks, you should only define the number of filters to be used and not the type of filters (for instance, a vertical edge detector). Each filter configuration (the numbers that it contains to detect a specific feature) is part of the variables of the system.

There are two additional concepts to be introduced to the subject of convolutional layers, which are as follows.

Padding

The padding parameter, as the name indicates, pads the image with zeros. This means that it adds additional pixels (which are filled with zeros) to each side of the image.

The following diagram shows an example of an image that has been padded by one on each side:

Figure 4.10: Graphical representation of an input image padded by one

This is used to maintain the shape of the input matrix once it has been passed through the filter. This is because, especially in the first couple of layers, the objective should be to preserve as much information from the original input as possible in order to extract the most features out of it.

To better understand the concept of padding, consider the following scenario.

Applying a 3 x 3 filter to a colored image of shape 32 x 32 x 3 would result in a matrix of shape 30 x 30 x 1. This means that the input for the following layer has shrunk. However, by adding a padding of 1 to the input image, the shape of the input is changed to 34 x 34 x 3, which results in an output of 32 x 32 x 1 using the same filter.

The following equation can be used to calculate the output width when using padding:

$$output\ width = W - F + 2 * P + 1$$

Figure 4.11: Output width after applying a convolutional layer using padding

Here, W refers to the width of the input matrix, F refers to the width of the filter, and P refers to the padding. The same equation can be adapted to calculate the height of the output.

To obtain an output matrix that's equal in shape to the input, use the following equation to calculate the value for the padding (considering that the stride, which we will define in the next section, is equal to one):

$$Padding = \frac{F - 1}{2}$$

Figure 4.12: Padding the number to get an output matrix that's equal in shape to the input

Keep in mind that the number of output channels (depth) will always be equal to the number of filters that have been applied to the input.

Stride

This parameter refers to the number of pixels that the filter will shift over the input matrix, both horizontally and vertically. As we have seen so far, the filter is passed through the top-left corner of the image, then it shifts over to the right by one pixel, and so on until it has gone through all the sections of the image vertically and horizontally. This example is one of a convolutional layer with a stride equal to one, which is the default configuration for this parameter.

When stride equals two, the shift would be two pixels instead, as seen in the following diagram:

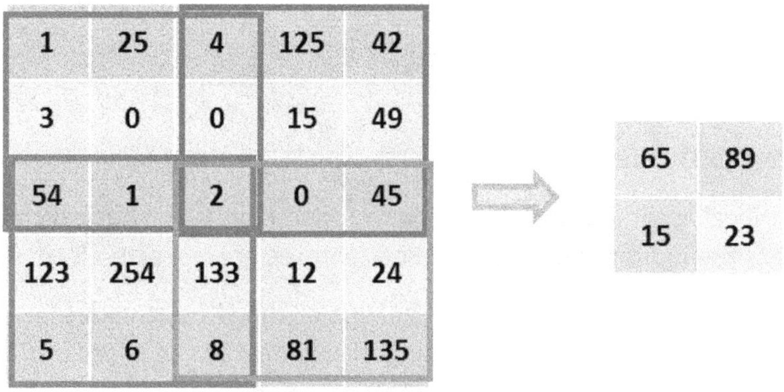

Figure 4.13: Graphical representation of a convolutional layer with a stride of two

As can be seen, the initial operation occurs in the top-left corner; then, by shifting two pixels to the right, the second calculation occurs in the top-right corner. Next, the calculation shifts two pixels downward to perform the calculations on the bottom-left corner, and, finally, by shifting two pixels to the right again, the final calculation occurs in the bottom-right corner.

> **NOTE**
>
> The numbers in *Figure 4.12* are made up, and not actual calculations. The focus should be on the boxes, which explain the shifting process when the stride is equal to two.

The following equation can be used to calculate the output width when using stride:

$$output\ width = \frac{(W-F)}{S} + 1$$

Figure 4.14: Output width of a convolutional layer using stride

Here, *W* refers to the width of the input matrix, *F* refers to the width of the filter, and *S* refers to the stride. The same equation can be adapted to calculate the height of the output.

Once these parameters have been introduced, the final equation to calculate the output shape (width and height) of the matrix that's been derived from a convolutional layer is as follows:

$$output\ width = \frac{(W - F) + 2 * P}{S} + 1$$

Figure 4.15: Output width after a convolutional layer using padding and stride

Whenever the value is a **float**, it should be rounded down. This basically means that some areas of the input are ignored, and no features are extracted from them.

Finally, once the input has been passed through all the filters, the output is fed to an activation function in order to break linearity, similar to the process of traditional neural networks. Although there are several activation functions that can be applied in this step, the preferred one is the ReLU function since it has shown outstanding results in CNNs. The output we obtained here becomes the input of the subsequent layer, which is usually a pooling layer.

EXERCISE 4.01: CALCULATING THE OUTPUT SHAPE OF A CONVOLUTIONAL LAYER

Using the equations given, consider the following scenarios and calculate the shape of the output matrix:

> **NOTE**
>
> This exercise does not require coding, but rather consists of calculations based on the concepts we mentioned previously.

1. Calculate the output shape of a matrix derived from a convolutional layer with an input of shape 64 x 64 x 3 and a filter of shape 3 x 3 x 3:

```
Output height = 64 -3 + 1 = 62
Output width = 64 - 3 + 1 = 62
Output depth = 1
```

2. Calculate the output shape of a matrix derived from a convolutional layer with an input of shape 32 x 32 x 3, 10 filters of shape 5 x 5 x 3, and padding of 2:

```
Output height = 32 - 5 + (2 * 2) + 1 = 32
Output width = 32-5 + (2 * 2) + 1 = 32
Output depth = 10
```

3. Calculate the output shape of a matrix derived from a convolutional layer with an input of shape 128 x 128 x 1, five filters of shape 5 x 5 x 1, and stride of 3:

```
Output height = (128 - 5)/ 3 + 1 = 42
Output width = (128 - 5)/ 3 + 1 = 42
Output depth = 5
```

4. Calculate the output shape of a matrix derived from a convolutional layer with an input of shape 64 x 64 x 1, a filter of shape 8 x 8 x 1, padding of 3, and a stride of 3:

```
Output height = ((64 - 8 + (2 * 3)) / 3) +1 = 21.6 ≈ 21
Output width = ((64 - 8 + (2 * 3)) / 3) +1 = 21.6 ≈ 21
Output depth = 1
```

With that, you have successfully calculated the output shape of the matrix that's been derived from a convolutional layer.

Coding a convolutional layer in PyTorch is very simple. Using custom modules, it only requires the creation of the **network** class. The class should contain an **__init__** method that defines the network architecture (that is, the layers of the network) and a **forward** method that defines the calculations to be performed on the information as it passes through the layers, as shown in the following code snippet:

```
import torch.nn as nn
import torch.nn.functional as F

class CNN_network(nn.Module):
    def __init__(self):
        super(CNN_network, self).__init__()
        self.conv1 = nn.Conv2d(3, 18, 3, 1, 1)

    def forward(self, x):
        x = F.relu(self.conv1(x))
        return x
```

When defining the convolutional layer, the arguments that are passed through from left to right refer to the input channels, output channels (number of filters), kernel size (filter size), stride, and padding.

The preceding example consists of a convolutional layer with three input channels, 18 filters, each of size 3, and stride and padding equal to 1.

Another valid approach, equivalent to the previous example, consists of a combination of the syntax from custom modules and the use of **Sequential** containers, as can be seen in the following code snippet:

```
import torch.nn as nn

class CNN_network(nn.Module):
    def __init__(self):
        super(CNN_network, self).__init__()
        self.conv1 = nn.Sequential(nn.Conv2d(3, 18, 3, 1, 1), \
                                   nn.ReLU())

    def forward(self, x):
        x = self.conv1(x)
        return x
```

Here, the definition of layers occurs inside the **Sequential** container. Typically, one container includes a convolutional layer, an activation function, and a pooling layer. A new set of layers is included in a different container below it.

In the preceding example, both the convolutional layer and the activation layer are defined within the **Sequential** container. Hence, in the **forward** method, there is no need to pass the output of the convolutional layer through the activation function as it has already been handled using the container.

POOLING LAYERS

Conventionally, pooling layers are the last part of the feature selection step, which is why a pooling layer can mostly be found after a convolutional layer. As we explained in the previous chapters, the idea is to extract the most relevant information out of subsections of the image. The size of the pooling layer is typically two, and the stride is equal to its size.

Pooling layers commonly reduce the input's height and width by half. This is important considering that, in order for convolutional layers to find all the features in an image, several filters need to be used, and the output from this operation can become too large, which means there are many parameters to consider. Pooling layers aim to reduce the number of parameters in the network by keeping the most relevant features. Selecting the relevant features out of subsections of the image occurs either by grabbing the maximum number or by averaging the numbers in that region.

For image classification tasks, it is most common to use max pooling layers over average pooling layers. This is because the former has shown better results in tasks where preserving the most relevant features is key, while the latter has been proven to work better in tasks such as smoothing images.

To calculate the shape of the output matrix, use the following equation:

$$output\ width = \frac{W - F}{S} + 1$$

Figure 4.16: Output matrix width after a pooling layer

Here, W refers to the width of the input, F refers to the size of the filter, and S refers to the stride. The same equation can be adapted to calculate the output height.

The channels or depth of the input remain unchanged as the pooling layer will perform the same operation on all the channels of the image. This means that the result from a pooling layer only affects the input in terms of width and length.

EXERCISE 4.02: CALCULATING THE OUTPUT SHAPE OF A SET OF CONVOLUTIONAL AND POOLING LAYERS

The following exercise will combine both convolutional and pooling layers. The objective is to determine the size of the output matrix after going through a set of layers.

> **NOTE**
>
> This exercise does not require coding, but rather consists of calculations based on the concepts we mentioned previously.

Consider the following sets of layers and specify the shape of the output layer at the end of all the transformations, considering an input image of size 256 x 256 x 3:

1. A convolutional layer with 16 filters of size 3, and stride and padding of 1.

2. A pooling layer with a filter of size two and stride of size two as well.

3. A convolutional layer with eight filters of size seven, stride of one, and padding of three.

4. A pooling layer with a filter of size two and a stride of two as well.

The output size of the matrix after going through each of these layers is as follows:

1. After the first convolutional layer:

   output_width/height = ((256 – 3) + 2 * 1)/1 + 1 = 256

   output_channels = 16 filters were applied

   output_matrix_size = 256 x 256 x 16

2. After the first pooling layer:

   output_width/height = (256 – 2) / 2 + 1 = 128

   output_channels = 16 as pooling does not affect the number of channels

   output_matrix_size = 128 x 128 x 16

3. After the second convolutional layer:

   output_width/height = ((128 – 7) + 2 =* 3)/1 + 1 = 128

   output_channels = 8 filters were applied

   output_matrix_size = 128 x 128 x 8

4. After the second pooling layer:

   output_width/height = (128 – 2) / 2 + 1 = 64

   output_channels = 8 as pooling does not affect the number of channels

   output_matrix_size = 64 x 64 x 8

With that, you have successfully calculated the output shapes of a matrix derived from a series of convolutional and pooling layers.

Using the same coding examples as before, the PyTorch way to define pooling layers is shown in the following code snippet:

```
import torch.nn as nn
import torch.nn.functional as F
class CNN_network(nn.Module):
    def __init__(self):
        super(CNN_network, self).__init__()
        self.conv1 = nn.Conv2d(3, 18, 3, 1, 1)
        self.pool1 = nn.MaxPool2d(2, 2)
```

```
    def forward(self, x):
        x = F.relu(self.conv1(x))
        x = self.pool1(x)
        return x
```

As can be seen, a pooling layer (**MaxPool2d**) was added to the network architecture in the **__init__** method. Here, the arguments that go into the max pooling layers, from left to right, are the size of the filter (**2**) and the stride (**2**). Next, the **forward** method was updated to pass the information through the new pooling layer.

Again, an equally valid approach is shown here, with the use of custom modules and **Sequential** containers:

```
import torch.nn as nn

class CNN_network(nn.Module):
    def __init__(self):
        super(CNN_network, self).__init__()
        self.conv1 = nn.Sequential(nn.Conv2d(3, 18, 3, 1, 1),\
                                   nn.ReLU(),\
                                   nn.MaxPool2d(2, 2))

    def forward(self, x):
        x = self.conv1(x)
        return x
```

As we mentioned previously, the pooling layer is also included in the same container as the convolutional layer, below the activation function. A subsequent set of layers (convolutional, activation, and pooling) would be defined below, in a new **Sequential** container.

Again, the **forward** method no longer needs to call each layer individually; instead, it passes the information through the container, which holds both the layers and the activation function.

FULLY CONNECTED LAYERS

The FC layer or layers are defined at the end of the network architecture after the input has gone through a set of convolutional and pooling layers. The output data from the layer preceding the first FC layer is flattened from a matrix into a vector, which can be fed to the FC layer (the same as a hidden layer from traditional neural networks).

The main purpose of these FC layers is to consider all the features that were detected by the previous layers, in order to classify the image.

The different FC layers are passed through an activation function, which is typically the ReLU function, unless it is the final layer, which will use a softmax function to output the probability of the input belonging to each of the class labels.

The input size of the first FC layer corresponds to the size of the flattened output matrix from the previous layer. The output size is defined by the user, and, again, as with ANNs, there is not an exact science to setting this number. The last FC layer should have an output size equal to the number of class labels.

To define a set of FC layers in PyTorch, consider the following code snippet:

```python
import torch.nn as nn
import torch.nn.functional as F


class CNN_network(nn.Module):

    def __init__(self):
        super(CNN_network, self).__init__()

        self.conv1 = nn.Conv2d(3, 18, 3, 1, 1)
        self.pool1 = nn.MaxPool2d(2, 2)

        self.linear1 = nn.Linear(32*32*16, 64)
        self.linear2 = nn.Linear(64, 10)


    def forward(self, x):
        x = F.relu(self.conv1(x))
        x = self.pool1(x)

        x = x.view(-1, 32 * 32 *16)
        x = F.relu(self.linear1(x))
        x = F.log_softmax(self.linear2(x), dim=1)

        return x
```

Using the same coding example as in the previous section, two FC layers are added to the network inside the __init__ method. Next, inside the **forward** function, the output from the pooling layer is flattened using the **view()** function. Then, it is passed through the first FC layer, which applies an activation function. Finally, the data is passed through a final FC layer, along with its activation function.

Again, using the same coding example as before, it is possible to add FC layers to our model using both custom modules and the **Sequential** container, as follows:

```python
import torch.nn as nn

class CNN_network(nn.Module):
    def __init__(self):
        super(CNN_network, self).__init__()
        self.conv1 = nn.Sequential(nn.Conv2d(1, 16, 5, 1, 2,), \
                                   nn.ReLU(), \
                                   nn.MaxPool2d(2, 2))

        self.linear1 = nn.Linear(32*32*16, 64)
        self.linear2 = nn.Linear(64, 10)

    def forward(self, x):
        x = self.conv1(x)

        x = x.view(-1, 32 * 32 *16)
        x = F.relu(self.linear1(x))
        x = F.log_softmax(self.linear2(x), dim=1)

        return x
```

As can be seen, the **Sequential** container is left untouched and the two FC layers are added below, inside the __init__ method. Next, the **forward** function passes the information through the entire container, to then flatten the output to be passed through the FC layers.

Once the architecture of the network has been defined, the following steps for training the network can be handled the same way as they are for ANNs.

SIDE NOTE – DOWNLOADING DATASETS FROM PYTORCH

To load a dataset from PyTorch, use the following code. Besides downloading the dataset, the following code shows how to use data loaders to save resources by loading the images in batches, rather than all at once:

```
from torchvision import datasets
import torchvision.transforms as transforms


transform = \
transforms.Compose([transforms.ToTensor(), \
                 transforms.Normalize((0.5, 0.5, 0.5), \
                                   (0.5, 0.5, 0.5))])
```

The **transform** variable is used to define the set of transformations to perform on the dataset. In this case, the dataset will be both converted into tensors and normalized in all its dimensions.

```
train_data = datasets.MNIST(root='data', train=True,\
                         download=True, transform=transform)


test_data = datasets.MNIST(root='data', train=False,\
                         download=True, transform=transform)
```

In the preceding code, the dataset to be downloaded is MNIST. This is a popular dataset that contains images of hand-written grayscale numbers from zero to nine. PyTorch datasets provide both training and testing sets.

As can be seen in the preceding snippet, to download the dataset, it is necessary to define the root of the data, which, by default, should be defined as **data**. Next, define whether you are downloading the training or the testing dataset. We set the **download** argument to **True**. Finally, we use the **transform** variable that we defined previously to perform the transformations on the datasets:

```
dev_size = 0.2
idx = list(range(len(train_data)))
np.random.shuffle(idx)
split_size = int(np.floor(dev_size * len(train_data)))
train_idx, dev_idx = idx[split_size:], idx[:split_size]
```

Considering that we need a third set of data (the validation set), the preceding code snippet is used to partition the training set into two sets. First, the size of the validation set is defined, and then the list of indexes that will be used for each of the datasets are defined (the training and the validation sets):

```
train_sampler = SubsetRandomSampler(train_idx)
dev_sampler = SubsetRandomSampler(dev_idx)
```

In the preceding snippet, the **SubsetRandomSampler()** function from PyTorch is used to divide the original training set into training and validation sets by randomly sampling indexes. This will be used in the following step to generate the batches that will be fed into the model in each iteration:

```
batch_size = 20
train_loader = torch.utils.data.DataLoader(train_data, \
                                           batch_size=batch_size, \
                                           sampler=train_sampler)
dev_loader = torch.utils.data.DataLoader(train_data, \
                                         batch_size=batch_size, \
                                         sampler=dev_sampler)
test_loader = torch.utils.data.DataLoader(test_data, \
                                          batch_size=batch_size)
```

The **DataLoader()** function is used to load the images in batches, for each of the sets of data. First, the variable containing the set is passed as an argument and then the batch size is defined. Finally, the samplers that we created in the preceding step are used to make sure that the batches that are used in each iteration are randomly created, which helps improve the performance of the model. The resulting variables (**train_loader**, **dev_loader**, and **test_loader**) of this function will contain the values for the features and the target separately.

> **NOTE**
>
> The more complex the problem and the deeper the network, the longer it takes for the model to train. Considering this, the activities in this chapter may take longer than the ones in previous chapters.

ACTIVITY 4.01: BUILDING A CNN FOR AN IMAGE CLASSIFICATION PROBLEM

In this activity, a CNN will be trained on an image dataset from PyTorch (that is, the framework provides the dataset). The dataset to be used is CIFAR10, which contains a total of 60,000 images of vehicles and animals. There are 10 different class labels (such as "airplane," "bird," "automobile," "cat," and so on). The training set contains 50,000 images, while the testing set contains the remaining 10,000.

> **NOTE**
>
> To explore this dataset even further, visit the following URL:
> https://www.cs.toronto.edu/~kriz/cifar.html.

Let's take a look at our scenario. You work at an artificial intelligence company that develops custom-made models for the needs of its customers. Your team is currently creating a model that can differentiate a picture of a vehicle from that of an animal and, more specifically, that can recognize different kinds of animals and different types of vehicles. They have provided you with a dataset containing 60,000 images to build the model.

> **NOTE**
>
> The activities in this chapter may take a long time to train on a regular computer (on a CPU). To run the code on a GPU, there is an equivalent file for each activity in this book's GitHub repository.

1. Import the required libraries.

2. Set the transformations to be performed on the data, which will be the conversion of the data into tensors and the normalization of the pixel values.

3. Set a batch size of 100 images and download both the training and testing data from the **CIFAR10** dataset.

4. Using a validation size of 20%, define the training and validation sampler that will be used to divide the dataset into those two sets.

5. Use the **DataLoader()** function to define the batches to be used for each setof data.

6. Define the architecture of your network. Use the following information to do so:

   Conv1: A convolutional layer that takes the colored image as input and passes it through 10 filters of size 3. Both the padding and the stride should be set to 1.

   Conv2: A convolutional layer that passes the input data through 20 filters of size 3. Both the padding and the stride should be set to 1.

   Conv3: A convolutional layer that passes the input data through 40 filters of size 3. Both the padding and the stride should be set to 1.

   Use the ReLU activation function after each convolutional layer.

   A pooling layer after each convolutional layer, with a filter size and stride of 2.

   A dropout term set to 20% after flattening the image.

   Linear1: A fully connected layer that receives the flattened matrix from the previous layer as input and generates an output of 100 units. Use the ReLU activation function for this layer. Here, the dropout term is set to 20%.

   Linear2: A fully connected layer that generates 10 outputs, one for each class label. Use the `log_softmax` activation function for the output layer.

7. Define all of the parameters that are required to train your model. Set the number of epochs to 50.

8. Train your network and be sure to save the values for the loss and accuracy of both the training and validation sets.

9. Plot the loss and accuracy of both sets.

10. Check the model's accuracy on the testing set—it should be around 72%.

> **NOTE**
>
> The solution to this activity can be found on page 262.
>
> Due to the data being shuffled in each epoch, the results will not be exactly reproducible. However, you should be able to arrive at similar results to the ones obtained in this book.
>
> This code may take time to run, which is why an equivalent GPU version solution is provided in this book's GitHub repository.

DATA AUGMENTATION

Learning how to effectively code a neural network is one of the steps involved in developing well-performing solutions. Additionally, to develop great deep learning solutions, it is crucial to find an area of interest in which we can provide a solution to a current challenge. But once all of that is done, we are typically faced with the same issue: getting a dataset of a decent size to get good performance from our models, either by self-gathering or by downloading it from the internet and other available sources.

As you might imagine, and even though it is now possible to gather and store vast amounts of data, this is not an easy task due to the costs associated with it. And so, most of the time, we are stuck working with a dataset containing tens of thousands of entries, and even fewer when referring to images.

This becomes a relevant issue when developing a solution for a computer vision problem, mainly due to two reasons:

- The larger the dataset, the better the results, and larger datasets are crucial to arrive at decent enough models. This is true considering that training a model is a matter of tuning a bunch of parameters so that it is capable of mapping a relationship between an input and an output. This is achieved by minimizing the loss function to make the predicted value come as close as possible to the ground truth. Here, the more complex the model, the more parameters it requires.

  Considering this, it is necessary to feed a fair number of examples to the model so that it is capable of finding such patterns, where the number of training examples should be proportional to the number of parameters to be tuned.

- One of the biggest challenges in computer vision problems is getting your model to perform well on several variations of an image. This means that images do not need to be fed by following a specific alignment or have a set quality, but can instead be fed in their original formats, including different positions, angles, lighting, and other distortions. Because of this, it is necessary to find a way to feed the model with such variations.

Due to this, the data augmentation technique was designed. Simply put, it is a measure that increases the number of training examples by slightly modifying the existing examples. For example, you could duplicate the instances currently available and add some noise to those duplicates to make sure they are not exactly the same.

In computer vision problems, this means incrementing the number of images in the training dataset by altering the existing images, which can be done by slightly altering the current images to create duplicated versions that are slightly different.

These minor adjustments to the images can be in the form of slight rotations, changes in the position of the object in the frame, horizontal or vertical flips, different color schemes, and distortions, among others. This technique works since CNNs will consider each of these images a different image.

For instance, the following image shows three images of a dog that, while to the human eye are the same image with certain variations, to the neural network are completely different:

Figure 4.17: Augmented images

A CNN that's capable of recognizing an object in an image regardless of a variation is considered to have the property of invariance. In fact, a CNN can be invariant to each type of variation.

DATA AUGMENTATION WITH PYTORCH

Performing data augmentation in PyTorch using the **torchvision** package is very easy. This package, in addition to containing popular datasets and model architectures, also contains common image transformation functions that can be performed on datasets.

> **NOTE**
>
> In this section, a few of these image transformations will be mentioned. To get the entire list of possible transformations, visit https://pytorch.org/docs/stable/torchvision/transforms.html.

As with the process we used in the previous activity to normalize and convert the dataset into tensors, performing data augmentation requires us to define the desired transformations, then apply them to the dataset, as shown in the following code snippet:

```
transform = transforms.Compose([\
            transforms.HorizontalFlip(probability_goes_here),\
            transforms.RandomGrayscale(probability_goes_here),\
            transforms.ToTensor(),\
            transforms.Normalize((0.5, 0.5, 0.5), \
                                 (0.5, 0.5, 0.5))])

train_data = datasets.CIFAR10('data', train=True, \
                              download=True, transform=transform)

test_data = datasets.CIFAR10('data', train=False, \
                             download=True, transform=transform)
```

Here, using the **HorizontalFlip** function, the data to be downloaded will undergo a horizontal flip (considering a probability value, which is set by the user and determines the percentage of images that will undergo this transformation). By using the **RandomGrayscale** function, the images will be converted into grayscale (also considering the probability). Then, the data is converted into tensors and normalized.

Considering that a model is trained in an iterative process, in which the training data is fed multiple times, these transformations ensure that a second run through the dataset does not feed the exact same images to the model.

Moreover, different transformations can be set for different sets. This is useful because the purpose of data augmentation is to increment the number of training examples, but the images that will be used for testing the model should be left mostly unaltered. Nevertheless, the testing set should be resized in order to feed equally sized images to the model.

This can be accomplished as follows:

```
transform = {"train": \
transforms.Compose([transforms.RandomHorizontalFlip\
                    (probability_goes_here),\
                    transforms.RandomGrayscale\
                    (probability_goes_here),\
                    transforms.ToTensor(),\
                    transforms.Normalize\
```

```
                     ((0.5, 0.5, 0.5), (0.5, 0.5, 0.5))]), \
           "test": transforms.Compose([transforms.ToTensor(),\
                                  transforms.Normalize\
                                    ((0.5, 0.5, 0.5), \
                                     (0.5, 0.5, 0.5)),\
           transforms.Resize(size_goes_here)])}

train_data = datasets.CIFAR10('data', train=True, download=True, \
            transform=transform["train"])

test_data = datasets.CIFAR10('data', train=False, download=True, \
            transform=transform["test"])
```

As we can see, a dictionary containing a set of transformations for the training and testing sets is defined. Then, the dictionary can be called to apply the transformations to each of the sets, accordingly.

ACTIVITY 4.02: IMPLEMENTING DATA AUGMENTATION

In this activity, data augmentation will be introduced to the model we created in the previous activity in order to test whether its accuracy can be improved. Let's look at the following scenario.

The model that you have created is good, but its accuracy is not at the desired level. You have been asked to think of a methodology that could improve the performance of the model. Follow these steps to complete this activity:

1. Duplicate the notebook from the previous activity.

2. Change the definition of the **transform** variable so that it includes, in addition to normalizing and converting the data into tensors, the following transformations:

    For the training/validation sets, a **RandomHorizontalFlip** function with a probability of 50% (0.5) and a **RandomGrayscale** function with a probability of 10% (0.1).

    For the testing set, do not add any other transformations.

3. Train the model for 100 epochs. The resulting plots for loss and accuracy on the training and validation sets should be similar to the ones shown here:

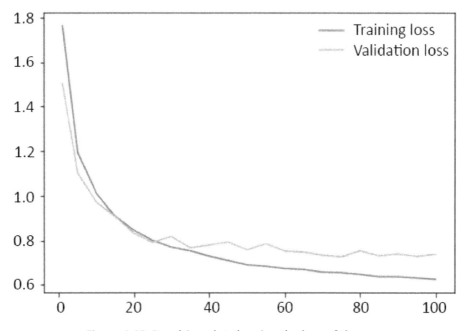

Figure 4.18: Resulting plot showing the loss of the sets

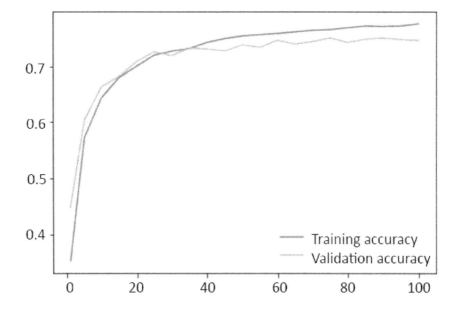

Figure 4.19: Resulting plot showing the accuracy of the sets

> **NOTE**
>
> Due to shuffling of the data in each epoch, the results will not be exactly reproducible. However, you should be able to arrive at similar results.

4. Calculate the accuracy of the resulting model on the testing set.

Expected output: The performance of the model on the testing set should be around 75%.

> **NOTE**
>
> The solution to this activity can be found on page 272.
>
> This code may take time to run, which is why an equivalent GPU version solution is provided in this book's GitHub repository.

BATCH NORMALIZATION

It is typical to normalize the input layer in an attempt to speed up learning, as well as to improve performance by rescaling all the features to the same scale. So, the question is, if the model benefits from the normalization of the input layer, why not normalize the output of all the hidden layers in an attempt to improve the training speed even more?

Batch normalization, as its name suggests, normalizes the outputs from the hidden layers so that it reduces the variance from each layer, which is also known as covariance shift. This reduction of the covariance shift is useful as it allows the model to also work well on images that follow a different distribution than the images used to train it.

Take, for instance, a network that has the purpose of detecting whether an animal is a cat. When the network is trained only using images of black cats, batch normalization can help the network also classify new images of cats of different colors by normalizing the data so that both the black and colored cat images follow a similar distribution. Such a problem is represented in the following image:

Figure 4.20: Cat classifier – the model can recognize colored cats, even after being trained using only black cats

In addition, batch normalization introduces the following benefits to the process of training the model, which ultimately helps you arrive at a better performing model:

- It allows a higher learning rate to be set as batch normalization helps to ensure that none of the outputs are too high or too low. A higher learning rate is equivalent to faster learning times.

- It helps reduce overfitting because it has a regularization effect. This makes it possible to set the dropout probability to a lower value, which means that less information is ignored in each forward pass.

NOTE

We should not rely mainly on batch normalization to deal with overfitting.

As we explained in previous sections, normalizing the output of a hidden layer is done by subtracting the batch mean and dividing by the batch standard deviation.

Furthermore, batch normalization is typically performed on the convolutional layers, as well as the FC layers (excluding the output layer).

BATCH NORMALIZATION WITH PYTORCH

In PyTorch, adding batch normalization is as simple as adding a new layer to the network architecture, considering that there are two different types, as explained here:

- **BatchNorm1d**: This layer is used to implement batch normalization on a two-dimensional or three-dimensional input. It receives the number of output nodes from the previous layer as an argument. This is commonly used on FC layers.

- **BatchNorm2d**: This applies batch normalization to four-dimensional inputs. Again, the argument that it takes is the number of output nodes from the previous layer. It is commonly used on convolutional layers, meaning that the argument that it takes in should be equal to the number of channels from the previous layer.

According to this, the implementation of batch normalization in a CNN is as follows:

```
class CNN(nn.Module):
    def __init__(self):
        super(CNN, self).__init__()
        self.conv1 = nn.Conv2d(3, 16, 3, 1, 1)
        self.norm1 = nn.BatchNorm2d(16)
        self.pool = nn.MaxPool2d(2, 2)
        self.linear1 = nn.Linear(16 * 16 * 16, 100)
        self.norm2 = nn.BatchNorm1d(100)
        self.linear2 = nn.Linear(100, 10)

    def forward(self, x):
        x = self.pool(self.norm1(F.relu(self.conv1(x))))
        x = x.view(-1, 16 * 16 * 16)
        x = self.norm2(F.relu(self.linear1(x)))
        x = F.log_softmax(self.linear2(x), dim=1)
        return x
```

As we can see, the batch normalization layers are initially defined in a similar way to any other layer inside the **__init__** method. Next, each batch normalization layer is applied to the output of its corresponding layer after the activation function inside the **forward** method.

ACTIVITY 4.03: IMPLEMENTING BATCH NORMALIZATION

For this activity, we will implement batch normalization on the architecture of the previous activity in order to see if it is possible to further improve the performance of the model on the testing set. Let's look at the following scenario.

You have impressed your teammates with the last improvement you made in terms of performance, and now they are expecting more from you. They have asked you to give improving the model one last try so that the accuracy goes up to 80%. Follow these steps to complete this activity:

1. Duplicate the notebook from the previous activity.

2. Add batch normalization to each convolutional layer, as well as to the first FC layer.

3. Train the model for 100 epochs. The resulting plots of the loss and accuracy of the training and validation sets should be similar to the ones shown here:

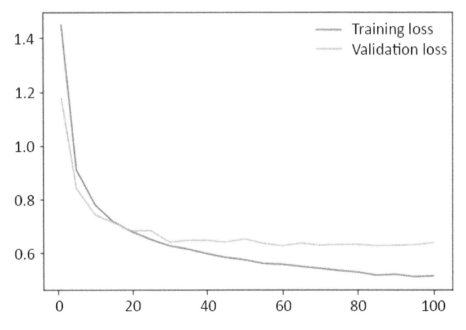

Figure 4.21: Resulting plot showing the loss of the sets

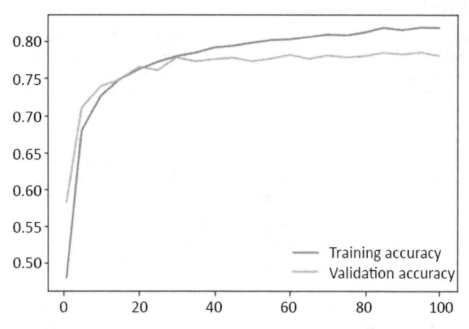

Figure 4.22: Resulting plot showing the loss of the sets

4. Calculate the accuracy of the resulting model on the testing set—it should be around 78%.

> **NOTE**
>
> The solution to this activity can be found on page 274.
>
> Due to shuffling the data in each epoch, the results will not be exactly reproducible. However, you should be able to arrive at similar results to the ones obtained in this book.
>
> This code may take time to run, which is why an equivalent GPU version solution is provided in this book's GitHub repository.

SUMMARY

This chapter focused on CNNs, which consist of a kind of neural network architecture that performs outstandingly well on computer vision problems. We started by explaining the main reasons why CNNs are widely used for dealing with image datasets, as well as providing an introduction to the different tasks that can be solved through their use.

This chapter explained the different building blocks of a network's architecture by explaining the nature of convolutional layers, pooling layers, and, finally, FC layers. In each section, an explanation of the purpose of each layer was included, as well as code snippets that can be used to effectively code the architecture in PyTorch.

This led to the introduction of an image classification problem focused on classifying images of vehicles and animals. The purpose of this problem was to put the different building blocks of CNNs into practice to solve an image classification data problem.

Next, data augmentation was introduced as a tool to improve a network's performance by incrementing the number of training examples, without the need to gather more images. This technique focuses on creating variations of the existing images to create "new" images to be fed to the model.

By implementing data augmentation, the second activity of this chapter aimed to solve the same image classification problem, with the objective of comparing results.

Finally, this chapter explained the concept of batch normalization. This consists of normalizing the output from each hidden layer in order to speed up learning. After explaining the process of applying batch normalization in PyTorch, the last activity of this chapter aimed to solve the same image classification problem using batch normalization.

Now that the concept of CNNs is clear and has been applied to solve a computer vision problem, in the next chapter, we will explore a more complex application of CNNs to create images, rather than just classifying them.

5

STYLE TRANSFER

OVERVIEW

This chapter explains the process of using pre-trained models in order to create or make use of well-performing algorithms without having to gather large quantities of data. In this chapter, you will learn how to load a pre-trained model from PyTorch in order to create a style transfer model. By the end of this chapter, you will be able to perform style transfer through the use of pretrained models.

INTRODUCTION

The previous chapter explained the different building blocks of traditional CNNs, as well as some techniques for improving their performance and reducing training time. The architecture explained there, although typical, is not set in stone, and a proliferation of CNN architectures have emerged to solve different data problems, most commonly in the field of computer vision.

These architectures vary in configuration as well as learning tasks. A very popular one nowadays is the **Visual Geometry Group** (**VGG**) architecture created by Karen Simonyan and Andrew Zisserman of Oxford's Robotic Institute. It was developed for object recognition and achieved state-of-the-art performance thanks to the massive number of parameters that the network relies on. One of the main reasons for its popularity among data scientists is the availability of the parameters (weights and biases) of the trained model, which allows researchers to use it without training, as well as the outstanding performance of the model.

In this chapter, we will use this pretrained model to solve a computer vision problem that is particularly famous due to the popularity of social media channels specializing in sharing images. It consists of performing style transfer in order to create a new image with the style (colors and textures) of one image and the content (shapes and objects) of another one.

This task is performed millions of times every day when applying filters over regular images to improve their quality and appeal while posting on social media profiles. Although it seems like a simple task, this chapter will explain the magic that occurs behind the scenes of these image editing apps.

> **NOTE**
>
> All the code present in this chapter can be found at:
> https://packt.live/2yiR97z.

STYLE TRANSFER

In simple words, style transfer consists of modifying the style of an image while still preserving its content. One example would be taking an image of an animal and transforming the style into a Monet-like painting, as shown in the following image:

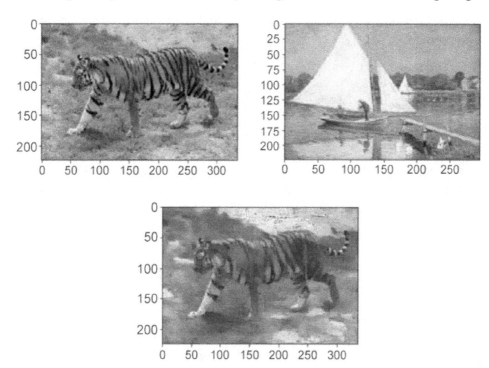

Figure 5.1: Style transfer inputs and output – the result of the final exercise of this chapter

NOTE

This image is available on GitHub at https://packt.live/2XEykpL.

According to the preceding image, there are two inputs to the model: a content image and a style image. Content refers to the objects of an image, while style refers to the colors and textures. As a result, the output from the model should be an image containing the objects from the content image and the artistic appearance of the style image.

HOW DOES IT WORK?

As opposed to solving a traditional computer vision problem, which was explained in the previous chapter, style transfer requires a different set of steps to effectively take two images as input and create a new image as the output.

The following is a brief explanation of the steps that are followed when solving a style transfer problem:

1. **Feeding the inputs**: Both the content and style images are to be fed to the model, and they need to be the same shape. A common practice here is to resize the style image so that it's the same shape as the content image.

2. **Loading the model**: Oxford's VGG created a model architecture that performs outstandingly well on style transfer problems, known as the VGG network. They also made the model's parameters available to anyone so that the training process of the model could be shortened or skipped (this is what a pretrained model is for).

   > **NOTE**
   >
   > There are different versions of the VGG network, and all use a different number of layers. To differentiate between the different versions, the nomenclature is such that a dash and a number at the end of the acronym represent the number of layers of that particular architecture. For this chapter, we will use the network's 19-layer version, which is known as VGG-19.

   Because of this, it is possible to load the pretrained model using PyTorch's `models` subpackage to perform the style transfer task without the need to train the network with a large number of images.

3. **Determining the layers' functions**: Given that there are two main tasks at hand (recognizing the content of an image and distinguishing the style of another image), different layers will have different functions to extract the different features. For the style image, the focus should be on colors and textures, while for the content image, the focus should be on edges and forms. In this step, the different layers are separated into different tasks.

4. **Defining the optimization problem**: As with any other supervised problem, it is necessary to define a loss function, which will have the responsibility of measuring the difference between the output and inputs. Unlike other supervised problems, the task of style transfer requires you to define three different loss functions, all of which should be minimized during the training process. The three loss functions are explained here:

   Content loss: This measures the distance between the content image and the output while only considering features related to content.

   Style loss: This measures the distance between the style image and the output while only considering features related to style.

   Total loss: This combines both the content and style loss. Both losses have a weight associated with them, which is used to determine their participation in the calculation of the total loss.

5. **Parameters update**: This step uses gradients to update the different parameters of the model.

IMPLEMENTATION OF STYLE TRANSFER USING THE VGG-19 NETWORK ARCHITECTURE

VGG-19 is a CNN consisting of 19 layers. It was trained using millions of images from the ImageNet database. The network is capable of classifying images into 1,000 different class labels, including a vast number of animals and different tools.

> **NOTE**
>
> To explore the ImageNet database, go to the following URL:
> http://www.image-net.org/.

Considering its depth, the network is able to identify complex features from a wide variety of images, which makes it particularly good for style transfer problems, where feature extraction is crucial at different stages and for different purposes.

This section will focus on how to use the pretrained VGG-19 model to perform style transfer. The end goal of this chapter will be to take an image of an animal or a landscape (as the content image) and one of a painting from a well-known artist (as the style image) to create a new image of a regular object with an artistic style.

However, before diving into this process, the following is a list of the imports and brief explanations of their use:

- **NumPy**: This will be used to transform the images to be displayed.

- **torch**, **torch.nn**, and **torch.optim**: These will implement the neural network and define the optimization algorithm.

- **PIL.Image**: This will load the images, as per the following code snippet:

```
image = Image.open(image_name)
image = transformation(image).unsqueeze(0)
```

As can be seen, the first step consists of opening the image (here, **image_name** should be replaced with the path to the image). Next, any transformation that was previously defined can be applied to the image.

> **NOTE**
>
> For a reminder of how to define transformations for images, please revisit *Chapter 4*, *Convolutional Neural Networks*.

The **unsqueeze()** function is used to add an extra dimension to the images, as per the requirements to feed images to the VGG-19 model.

- **matplotlib.pyplot**: This will display images.

- **torchvision.transforms** and **torchvision.models**: These will convert the images into tensors and load the pretrained model.

INPUTS – LOADING AND DISPLAYING

The first step of performing style transfer consists of loading both the content and style images. During this step, the basic pre-processing is handled, where images must be of equal size (preferably the size of the images used to train the pre-trained model), which will be the size of the output image as well. Additionally, images are converted into PyTorch tensors and can be normalized if desired.

It is always good practice to display the images that have been loaded in order to make sure that they are as desired. Considering that the images have already been converted into tensors and normalized at this point, the tensor should be cloned, and a new set of transformations need to be performed so that we can display them using Matplotlib. This means that the tensors should be converted back into **Python Imaging Library (PIL)** images and the normalization process must be reverted, as in the following example:

```
image = tensor.clone()
image = image.squeeze(0)

img_display = \
transforms.Compose([transforms.Normalize((-0.5/0.25, \
                                          -0.5/0.25, -0.5/0.25), \
                                          (1/0.25, 1/0.25, \
                                          1/0.25)), \
                transforms.ToPILImage()])
```

First, the tensor is cloned, and the additional dimension is removed. Next, the transformations are defined.

To understand the process of reverting the normalization, consider an image normalized using, for all dimensions, 0.5 as the mean and 0.25 as the standard deviation. The way to revert the normalization would be to use the negative value of the mean divided by the standard deviation as the mean (-0.5 divided by 0.25). The new standard deviation should be equal to one divided by the standard deviation (1 divided by 0.25). Defining functions to both load and display images can help save time and make sure that the same process is done over both the content and style images. This process will be expanded on in the following exercise.

> **NOTE**
>
> All the exercises for this chapter are to be coded in the same notebook as, together, the combined code will perform the style transfer task.

EXERCISE 5.01: LOADING AND DISPLAYING IMAGES

This is the first of four steps of performing style transfer. The objective of this exercise is to load and display the images (both content and style images) that will be used in further exercises. Follow these steps to complete this exercise:

> **NOTE**
>
> For the exercises and activities in this chapter, you will need to have Python 3.7, Jupyter 6.0, Matplotlib 3.1, NumPy 1.17, Pillow 6.2, and PyTorch 1.3+ (preferably PyTorch 1.4, with or without CUDA) installed (as instructed in the *Preface*).
>
> Inside this book's GitHub repository (https://packt.live/2yiR97z), you will be able to find different images that will be used throughout this chapter in the different exercises and activities.

1. Import all the packages that will be required to perform style transfer:

```
import numpy as np
import torch
from torch import nn, optim
from PIL import Image
import matplotlib.pyplot as plt
from torchvision import transforms, models
```

If you have a GPU available, define a variable named **device** equal to **cuda**, which will be used to allocate some variables to the GPU of your machine:

```
device = "cuda"
device
```

2. Set the image size to be used for both images. Also, set the transformations to be performed over the images, which should include resizing the images, converting them into tensors, and normalizing them:

```
imsize = 224

loader = transforms.Compose([\
        transforms.Resize(imsize), \
        transforms.ToTensor(), \
        transforms.Normalize((0.485, 0.456, 0.406), \
                             (0.229, 0.224, 0.225))])
```

Using this code, the images are resized to the same size as the images that were originally used to train the VGG-19 model. Normalization is done using the same values that were used to normalize the training images as well.

> **NOTE**
>
> The VGG network was trained using normalized images, where each channel has a mean of 0.485, 0.456, and 0.406, respectively, and a standard deviation of 0.229, 0.224, and 0.225, respectively.

3. Define a function that will receive the image path as input and use **PIL** to open the image. Next, it should apply the transformations to the image:

```
def image_loader(image_name):
    image = Image.open(image_name)
    image = loader(image).unsqueeze(0)
    return image
```

4. Call the function to load the content and style images. Use the dog image as the content image and the Matisse image as the style image, both of which are available in this book's GitHub repository:

```
content_img = image_loader("images/dog.jpg")
style_img = image_loader("images/matisse.jpg")
```

If your machine has a GPU available, use the following code snippet instead to achieve the same results:

```
content_img = image_loader("images/dog.jpg").to(device)
style_img = image_loader("images/matisse.jpg").to(device)
```

The preceding code snippet allocates the variables holding the images to the GPU so that all the operations using these variables are handled by the GPU.

5. To display the images, convert them back into PIL images and revert the normalization process. Define these transformations in a variable:

```
unloader = transforms.Compose([\
            transforms.Normalize((-0.485/0.229, \
                                  -0.456/0.224, \
                                  -0.406/0.225), \
                                 (1/0.229, 1/0.224, 1/0.225)),\
            transforms.ToPILImage()])
```

6. Create a function that clones the tensor, squeezes it, and applies the transformations defined in the previous step to the tensor:

```
def tensor2image(tensor):

    image = tensor.clone()
    image = image.squeeze(0)
    image = unloader(image)

    return image
```

If your machine has a GPU available, use the following equivalent code snippet instead:

```
def tensor2image(tensor):
    image = tensor.to('cpu').clone()
    image = image.squeeze(0)
    image = unloader(image)
    return image
```

The preceding code snippet allocates the images back to the CPU so that we can plot them.

7. Call the function for both images and plot the results:

```
plt.figure()
plt.imshow(tensor2image(content_img))
plt.title("Content Image")
plt.show()

plt.figure()
plt.imshow(tensor2image(style_img))
plt.title("Style Image")
plt.show()
```

The resulting images should look as follows:

Figure 5.2: Content image

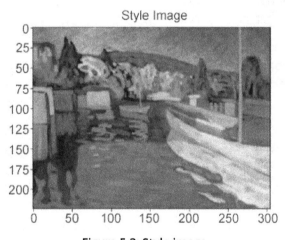

Figure 5.3: Style image

With that, you have successfully loaded and displayed the content and style images to be used for style transfer.

LOADING THE MODEL

As in many other frameworks, PyTorch has a subpackage that contains different models that have been previously trained and made available for public use. This is important considering that training a neural network from scratch is time-consuming; starting off with a pre-trained model can help reduce this training time. This means that pre-trained models can be loaded so that we can use their final parameters (which should be those that minimize the loss function) without the need to go through an iterative process.

As we mentioned earlier, the architecture that's used to perform the style transfer task is that of the VGG network of 19 layers, also known as VGG-19. The pre-trained model is available under the **models** subpackage of **torchvision**. The saved model in PyTorch is split into two portions:

1. **vgg19.features**: This consists of all the convolutional and pooling layers of the network, along with the parameters. These layers are in charge of extracting the features from the images; while some of the layers specialize in style features, such as colors, others specialize in content features, such as edges.

2. **vgg19.classifier**: This refers to the linear layers (also known as fully connected layers) that are located at the end of the network, including their parameters. These layers are the ones that perform the classification of the image into one of the label classes, for instance, recognizing the type of animal in an image.

> **NOTE**
>
> To explore the other pre-trained models that are available in PyTorch, visit https://pytorch.org/docs/stable/torchvision/models.html.

According to the preceding information, only the features portion of the model should be loaded in order to extract the necessary features of the content and style images. Loading a model consists of calling the **models** subpackage, followed by the name of the model, making sure that the **pretrained** argument is set to **True** (in order to load the parameters from a previous training process), and that only the feature layers are being loaded, as per the following snippet:

```
model = models.vgg19(pretrained=True).features
```

The parameters (weights and biases) in each layer should be kept unchanged, considering that those are the ones that will help detect the desired features. This can be achieved by defining that the model does not need to calculate gradients for any of these layers, as follows:

```
for param in model.parameters():

    param.requires_grad_(False)
```

Here, for every parameter of the model that was loaded previously, the **requires_ grad_** method is set to **False** in order to avoid calculating gradients since the objective is to make use of the pre-trained parameters, without updating them.

EXERCISE 5.02: LOADING A PRE-TRAINED MODEL IN PYTORCH

Using the same notebook as in the previous exercise, this exercise aims to load the pre-trained model that will be used in subsequent exercises to perform the style transfer task using the images we loaded previously.

1. Open the notebook from the previous exercise.

2. Load the VGG-19 pre-trained model from PyTorch:

```
model = models.vgg19(pretrained=True).features
```

   Select the features portion of the model, as explained previously. This will give you access to all the convolutional and pooling layers of the model, which are to be used to perform the extraction of features in subsequent exercises of this chapter.

3. Perform a **for** loop through the parameters of the previously loaded model. Set each parameter so that it doesn't require gradients calculations:

```
for param in model.parameters():
    param.requires_grad_(False)
```

By setting the calculation of gradients to **False**, we ensure that no gradients are calculated during the training process.

If your machine has a GPU available, add the following code snippet to the preceding snippet in order to allocate the model to the GPU:

```
model.to(device)
```

> **NOTE**
>
> To access the source code for this specific section, please refer to https://packt.live/2VCYqla.
>
> You can also run this example online at https://packt.live/2BXLXYE. You must execute the entire Notebook in order to get the desired result.
>
> To access the GPU version of this source code, please refer to https://packt.live/2Vx2kC4. This version of the source code is not available as an online interactive example, and will need to be run locally with the GPU setup.

With that, you have successfully loaded a pre-trained model.

EXTRACTING THE FEATURES

The VGG-19 network, as we mentioned previously, contains 19 different layers, including convolutional, pooling, and fully connected layers. The convolutional layers come in stacks before every pooling layer, with five being the number of stacks in the entire architecture.

In the field of style transfer, there have been different papers that have identified those layers that are crucial for recognizing relevant features over the content and style images. Accordingly, it is conventionally accepted that the first convolutional layer of every stack is capable of extracting style features, while only the second convolutional layer of the fourth stack should be used to extract content features.

From now on, we will refer to the layers that extract the style features as **conv1_1**, **conv2_1**, **conv3_1**, **conv4_1**, and **conv5_1**, while the layer in charge of extracting the content features will be known as **conv4_2**.

> **NOTE**
>
> The paper that was used as a guide for this chapter can be accessed at the following URL: https://www.cv-foundation.org/openaccess/content_cvpr_2016/papers/Gatys_Image_Style_Transfer_CVPR_2016_paper.pdf.

This means that the features of the style image are obtained from five different layers, while the features of the content image are only obtained from one layer. The output from each of these layers is used to compare the output image with the input images, where the objective would be to modify the parameters of the target image so that they resemble the content of the content image and the style of the style image, which can be achieved by optimizing three different loss functions (which will be further explained in this chapter).

To extract the features of each layer, the following code snippet can be used:

```
layers = {'0': 'conv1_1', '5': 'conv2_1', '10': 'conv3_1', \
          '19': 'conv4_1', '21': 'conv4_2', '28': 'conv5_1'}

features = {}
x = image

for index, layer in model._modules.items():
    x = layer(image)
    if index in layers:
        features[layers[index]] = x
```

In the preceding snippet, **layers** is a dictionary that maps the position (in the network) of all the relevant layers to the names that will be used to recognize them, and **model._modules** contains a dictionary holding each layer of the network.

By performing a **for** loop through the different layers, we pass the image through the different layers and save the output from the layers of interest (the ones inside the **layers** dictionary we created previously) into the **features** dictionary. The output dictionary consists of keys containing the name of the layer and values containing the output features from that layer.

To determine whether the target image contains the same content as the content image, we need to check whether certain features are present in both images. However, to check the style representation of the target image and the style image, it is necessary to check for correlations and not the strict presence of the features of both images. This is because the style features of both images will not be exact, but rather an approximation.

The gram matrix is used to check these correlations. It consists of the creation of a matrix that looks at the correlations of different style features in a given layer. This is done by multiplying the vectorized output from the convolutional layer by the same transposed vectorized output, as can be seen in the following diagram:

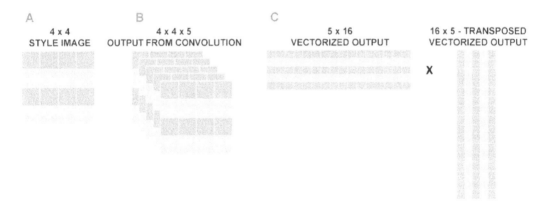

Figure 5.4: Calculation of the gram matrix

In the preceding diagram, A refers to the input style image with 4x4 dimensions (height and width), while B represents the output after passing the image through a convolutional layer with five filters. Finally, C refers to the calculation of the gram matrix, where the image to the left represents the vectorized version of B, and the image to the right is its transposed version. From the multiplication of the vectorized outputs, a 5x5 gram matrix is created, whose values indicate the similarities (correlations) in terms of style features along the different channels (filters).

These correlations can be used to determine those features that are relevant for the style representation of the image, which can then be used to alter the target image. Considering that the style features are obtained in five different layers, it is safe to assume that the network is capable of detecting small and large features from the style image, considering that a gram matrix has to be created for each of the layers.

EXERCISE 5.03: SETTING UP THE FEATURE EXTRACTION PROCESS

Using the network architecture from the previous exercise and the images from the first exercise of this chapter, we will create a couple of functions capable of extracting features from the input images and creating the gram matrix for the style features.

1. Open the notebook from the previous exercise.

2. Print the architecture of the model we loaded in the previous exercise. This will help us identify the relevant layers so that we can perform the style transfer task:

```
print(model)
```

3. Create a dictionary for mapping the index of the relevant layers (keys) to a name (values). This will facilitate the process of calling relevant layers in the future:

```
relevant_layers = {'0': 'conv1_1', '5': 'conv2_1', '10': \
                    'conv3_1', '19': 'conv4_1', '21': \
                    'conv4_2', '28': 'conv5_1'}
```

To create the dictionary, we use the output from the previous step, which displays each of the layers in the network. There, it is possible to observe that the first layer of the first stack is labeled **0**, while the first layer of the second stack is labeled **5**, and so on.

4. Create a function that will extract the relevant features (features extracted from the relevant layers only) from an input image. Name it **features_extractor** and make sure it takes the image, the model, and the dictionary we created previously as inputs:

```
def features_extractor(x, model, layers):

    features = {}
    for index, layer in model._modules.items():
        x = layer(x)
```

```
    if index in layers:
        features[layers[index]] = x

    return features
```

The output should be a dictionary, with the keys being the name of the layer and the values being the output features from that layer.

5. Call the **features_extractor** function over the content and style images we loaded in the first exercise of this chapter:

```
content_features = features_extractor(content_img, model, \
                                      relevant_layers)
style_features = features_extractor(style_img, model, \
                                    relevant_layers)
```

6. Perform the gram matrix calculation over the style features. Consider that the style features were obtained from different layers, which is why different gram matrices should be created, one for each layer's output:

```
style_grams = {}
for i in style_features:
    layer = style_features[i]
    _, d1, d2, d3 = layer.shape
    features = layer.view(d1, d2 * d3)
    gram = torch.mm(features, features.t())
    style_grams[i] = gram
```

For each layer, the shape of the style features' matrix is obtained in order to vectorize it. Next, the gram matrix is created by multiplying the vectorized output by its transposed version.

7. Create an initial target image. This image will be compared against the content and style images later and be changed until the desired similarity is achieved:

```
target_img = content_img.clone().requires_grad_(True)
```

It is good practice to create the initial target image as a copy of the content image. Moreover, it is essential to set it to require the calculation of gradients, considering that we want to be able to modify it in an iterative process until the content is similar to that of the content image and the style to that of the style image.

Again, if your machine has a GPU available, make sure that you allocate the target image to the GPU as well, using the following code snippet instead:

```
target_img = content_img.clone().requires_grad_(True).to(device)
```

8. Using the **tensor2image** function we created during the first exercise of this chapter, plot the target image, which should look the same as the content image:

```
plt.figure()
plt.imshow(tensor2image(target_img))
plt.title("Target Image")
plt.show()
```

The output image is as follows:

Figure 5.5: The target image

> **NOTE**
>
> To access the source code for this specific section, please refer to https://packt.live/2ZtSoL7.
>
> You can also run this example online at https://packt.live/2Vz7Cgm. You must execute the entire Notebook in order to get the desired result.
>
> To access the GPU version of this source code, please refer to https://packt. live/3ePLxIA. This version of the source code is not available as an online interactive example, and will need to be run locally with the GPU setup.

With that, you have successfully performed feature extraction and calculated the gram matrix to perform style transfer.

THE OPTIMIZATION ALGORITHM, LOSSES, AND PARAMETER UPDATE

Although style transfer is performed using a pre-trained network where the parameters are left unchanged, creating the target image consists of an iterative process where three different loss functions are calculated and minimized by updating only the parameters related to the target image.

To achieve the creation of the target image, two different loss functions are calculated (the content and style losses), which are then put together to calculate a total loss function that is to be optimized to arrive at an appropriate target image. However, considering that measuring accuracy in terms of content and style is achieved very differently, the following is an explanation of the calculation of both the content and style loss functions, as well as a description of how the total loss is calculated.

CONTENT LOSS

This consists of a function that, based on the feature map obtained by a given layer, calculates the distance between the content image and the target image. In the case of the VGG-19 network, the content loss is only calculated based on the output from the **conv4_2** layer.

The main idea behind the content loss function is to minimize the distance between the content image and the target image so that the latter highly resembles the former in terms of content.

The content loss can be calculated as the mean squared difference between the feature maps of the content and target images at the relevant layer (**conv4_2**), which can be achieved using the following equation:

$$content\ loss = torch.mean\big(\big(target\ features - content\ features\big)**2\big)$$

Figure 5.6: The content loss function

STYLE LOSS

Similar to the content loss, the style loss is a function that measures the distance between the style and target images in terms of style features (for instance, color and texture) by calculating the mean squared difference.

Contrary to the case of content loss, instead of comparing the feature maps derived from the different layers, it compares the gram matrices calculated based on the feature maps of both the style and target images.

The style loss has to be calculated for all the relevant layers (in this case, five layers) using a **for** loop. This will result in a loss function that considers simple and complex style representations from both images.

Furthermore, it is good practice to weigh the style representation of each of these layers between zero to one in order to give more emphasis to the layers that extract larger and simpler features over layers that extract very complex features. This is achieved by giving higher weights to earlier layers (**conv1_1** and **conv2_1**) that extract more generic features from the style image.

The calculation of the style loss can be performed using the following equation for each of the relevant layers:

$$style\ loss = style\ layer\ weight * torch.mean((target\ gram - style\ gram) * *2)$$

Figure 5.7: Style loss calculation

TOTAL LOSS

Finally, the total loss function consists of a combination of both the content loss and style loss. Its value is minimized during the iterative process of creating the target image by updating the parameters of the target image.

Again, it is recommended that you assign weights to the content and the style losses in order to determine their participation in the final output. This helps determine the degree to which the target image will be stylized while leaving the content still visible. It is good practice to set the weight of the content loss equal to one, whereas the weight for the style loss must be much higher to achieve the ratio of your preference.

The weight that's assigned to the content loss is conventionally known as alpha, while the one given to the style loss is known as beta.

The final equation to calculate the total loss is as follows:

$$total\ loss = content\ loss * alpha + style\ loss * beta$$

Figure 5.8: Total loss calculation

Once the weights of the losses have been defined, it is time to set the number of iteration steps, as well as the optimization algorithm, which should only affect the target image. This means that, in every iteration step, all three losses will be calculated so that we can use the gradients to optimize the parameters associated with the target image, until the loss functions are minimized and a target image with the desired look is achieved.

As with the optimization of previous neural networks, the following are the steps that can be observed in each iteration:

1. Get the features, both in terms of content and style, from the target image. In the initial iteration, this image will be an exact copy of the content image.

2. Calculate the content loss. This is done by comparing the content features map of the content and target images.

3. Calculate the average style loss of all relevant layers. This is achieved by comparing the gram matrices for all the layers of both the style and target images.

4. Calculate the total loss.

5. Calculate the partial derivatives of the total loss function with respect to the parameters (weights and biases) of the target image.

6. Repeat this until the desired number of iterations has been reached.

The final output will be an image with content similar to that of the content image, and a style similar to that of the style image.

EXERCISE 5.04: CREATING THE TARGET IMAGE

In the final exercise of this chapter, you will perform the task of style transfer. This exercise consists of coding the section in charge of performing the different iterations while optimizing the loss functions in order to arrive at an ideal target image. To do so, it is crucial to make use of the code bits we programmed in the previous exercises of this chapter:

NOTE

When running this code on a GPU, some alterations apply. Please visit this book's GitHub repository to revise the GPU version of this code.

1. Open the notebook from the previous exercise.

2. Define a dictionary containing the weights for each of the layers in charge of extracting style features:

```
style_weights = {'conv1_1': 1., 'conv2_1': 0.8, 'conv3_1': 0.6, \
                 'conv4_1': 0.4, 'conv5_1': 0.2}
```

Be sure to use the same names that you gave your layers in the previous exercise as keys.

3. Define the weights associated with the content and style losses:

```
alpha = 1
beta = 1e5
```

4. Define the number of iteration steps, as well as the optimization algorithm. We can also set the number of iterations if we want to see a plot of the image that has been created at that point:

```
print_statement = 200
optimizer = torch.optim.Adam([target_img], lr=0.001)
iterations = 2000
```

The parameters to be updated by this optimization algorithm should be the parameters of the target image.

> **NOTE**
>
> Running 2,000 iterations, as in the example in this exercise, will take quite some time, depending on your resources. However, to reach an outstanding target image, even more iterations may be required.
>
> For you to appreciate the changes that occur to the target image in each iteration, a couple of iterations will suffice, but you are encouraged to try training for longer.

5. Define the **for** loop where all three loss functions will be calculated and the optimization process will be performed:

```python
for i in range(1, iterations+1):
    # Extract features for all relevant layers
    target_features = features_extractor(target_img, model, \
                                            relevant_layers)
    # Calculate the content loss
    content_loss = torch.mean((target_features['conv4_2'] \
                                - content_features['conv4_2'])**2)

    # Loop through all style layers
    style_losses = 0
    for layer in style_weights:

        # Create gram matrix for that layer
        target_feature = target_features[layer]
        _, d1, d2, d3 = target_feature.shape

        target_reshaped = target_feature.view(d1, d2 * d3)
        target_gram = torch.mm(target_reshaped, \
                            target_reshaped.t())
        style_gram = style_grams[layer]

        # Calculate style loss for that layer
        style_loss = style_weights[layer] \
                    * torch.mean((target_gram - style_gram)**2)

        #Calculate style loss for all layers
        style_losses += style_loss / (d1 * d2 * d3)

    # Calculate the total loss
    total_loss = alpha * content_loss + beta * style_losses

    # Perform back propagation
    optimizer.zero_grad()
```

```
    total_loss.backward()
    optimizer.step()

    # Print target image
    if  i % print_statement == 0 or i == 1:
        print('Total loss: ', total_loss.item())
        plt.imshow(tensor2image(target_img))
        plt.show()
```

6. Plot both the content and target images to compare the results. This can be achieved by using the **tensor2image** function, which we created in the previous exercises, in order to convert the tensors into PIL images that can be printed using **matplotlib**:

```
fig, (ax1, ax2, ax3) = plt.subplots(1, 3, figsize=(20, 10))
ax1.imshow(tensor2image(content_img))
ax2.imshow(tensor2image(target_img))
ax3.imshow(tensor2image(style_img))
plt.show()
```

The final image should look similar to the following:

Figure 5.9: Comparison between the content, style, and target image

> **NOTE**
>
> To view the high-quality color image, visit this book's GitHub repository at
> https://packt.live/2VBZA5E.

With that, you have successfully performed style transfer.

> **NOTE**
>
> To access the source code for this specific section, please refer to https://packt.live/2VyKJtK.
>
> This section does not currently have an online interactive example, and will need to be run locally.
>
> To access the GPU version of this source code, please refer to https://packt.live/2YMcdhh. This version of the source code is not available as an online interactive example, and will need to be run locally with the GPU setup.

ACTIVITY 5.01: PERFORMING STYLE TRANSFER

In this activity, we will perform a style transfer. To do so, we will code all the concepts we've learned about throughout this chapter. Let's look at the following scenario.

You have some images that you want to alter so that they have an artistic flair and, to achieve that, you have decided to create some code that uses a pre-trained neural network to perform style transfer. Follow these steps to complete this activity:

1. Import the required libraries.

2. Specify the transformations to be performed over the input images. Be sure to resize them to the same size, convert them into tensors, and normalize them.

3. Define an image loader function. This should open the image and transform it. Call the image loader function to load both input images.

4. To be able to display the images, define a new set of transformations to revert the normalization of the images and to convert the tensors into PIL images.

5. Create a function (**tensor2image**) that's capable of performing the previous transformation over the tensors. Call the function for both images and plot the results.

6. Load the VGG-19 model.

7. Create a dictionary that maps the index of the relevant layers (keys) to a name (values). Then, create a function to extract the feature maps of the relevant layers. Use them to extract the features of both input images.

8. Calculate the gram matrix for the style features. Also, create the initial target image.

9. Set the weights for the different style layers, as well as the weights for the content and style losses.

10. Run the model for 500 iterations. Define the Adam optimization algorithm before starting to train the model, using 0.001 as the learning rate.

> **NOTE**
>
> Depending on your resources, the training process may take several hours. As such, to achieve excellent results, it is recommended that you train for thousands of iterations. Adding print statements is good practice if you wish to see the progress of the training process.
>
> The results shown in this chapter were achieved by running around 5,000 iterations, which will take a long time to run without a GPU (the solution to this activity where we used a GPU can also be found in this book's GitHub repository). However, to just see some minor changes, it will suffice to run it for a few hundred iterations, as recommended in this activity (500).

11. Plot the content, style, and target images to compare the results.

    The output after 5,000 iterations should appear as follows:

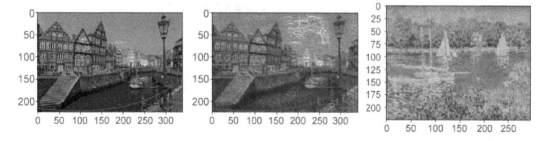

Figure 5.10: Plotting of the content and target images

> **NOTE**
>
> The solution to this activity can be found on page 277.
>
> To view the high-quality color image of *Figure 5.10*, visit https://packt.live/2KcORcw.

SUMMARY

This chapter introduced style transfer, which is a popular task nowadays that can be performed using CNNs. It consists of taking both a content image and a style image as inputs and returning a newly created image as output that keeps the content of one of the images and the style of the other. It is typically used to give images an artistic look by combining random regular images with those of the paintings of great artists.

Although style transfer is performed using CNNs, the process of creating the target image is not achieved by training the network conventionally. This chapter explained how to use pre-trained networks to consider the output of some relevant layers that are especially good at identifying certain features.

This chapter explained each of the steps required to develop code that's capable of performing the task of style transfer, where the first step consisted of loading and displaying the inputs. As we mentioned earlier, there are two inputs to the model (the content and style images). Each image is to go through a series of transformations, the aim being to resize the images to an equal size, convert them into tensors, and normalize them in order for them to be properly processed by the network.

Next, the pre-trained model was loaded. As we mentioned in this chapter, VGG-19 is one of the most commonly used architectures to solve such tasks. It consists of 19 layers, including convolutional, pooling, and fully connected layers, where, for the task in question, only some of the convolutional layers are to be used. The process of loading the pre-trained model is fairly simple, considering that PyTorch provides a subpackage containing several pre-trained network architectures.

Once the network is loaded, certain layers of the network are identified as overperformers at detecting certain features that are crucial for style transfer. While five different layers have the ability to extract features related to the style of an image, such as colors and textures, just one of the layers is exceptionally good at extracting content features, such as edges and shapes. Accordingly, it is crucial to define those layers that will be used to extract the information from the input images in order to create the desired target image.

Finally, it was time to code the iterative process to be used to create a target image with the desired features. To do so, three different losses were calculated. There was one for comparing the difference between the content image and the target image in terms of content (content loss), and another one for comparing the difference in terms of style between the style image and the target image (the style loss), which is achieved by the calculation of the gram matrix. Lastly, there was one that combined both the content and style losses (the total loss).

The target image is created by minimizing the value of the total loss, which can be done by updating the parameters related to the target image. Although a pre-trained network can be used, the process of arriving at an ideal target image may take several thousand iterations and quite some time.

In the next chapter, a different network architecture will be explained in order to solve a data problem using a sequence of text data. RNNs are neural network architectures that hold memory, which allows them to process sequential data. They are typically used to solve problems related to the understanding of human language.

6

ANALYZING THE SEQUENCE OF DATA WITH RNNS

OVERVIEW

This chapter expands on the concept of recurrent neural networks. You will learn about the learning process of **Recurrent Neural Networks (RNNs)** and how they store memory. The chapter will introduce the **Long Short-Term Memory (LSTM)** network architecture, which uses short- and long-term memory to solve data problems using sequences of data. By the end of this chapter, you will have a firm grasp of RNNs and of how to solve **Natural Language Processing (NLP)** data problems.

INTRODUCTION

In the previous chapters, different network architectures were explained – from traditional ANNs, which can solve both classification and regression problems, to CNNs, which are mainly used to solve computer vision problems by performing the tasks of object classification, localization, detection, and segmentation.

In this final chapter, we will explore the concept of RNNs and solve sequential data problems. These network architectures are capable of handling sequential data where context is crucial, thanks to their ability to hold information from previous predictions, which is called memory. This means that, for instance, when analyzing a sentence word by word, RNNs have the ability to hold information about the first word of the sentence when they are handling the last one.

This chapter will explore the LSTM network architecture, which is a type of RNN that can hold both long-term and short-term memory and is especially useful for handling long sequences of data, such as video clips.

The chapter will also explore the concept of NLP. NLP refers to the interaction of computers with human languages, which is a popular topic nowadays thanks to the rise of virtual assistants, which provide customized customer services. This chapter will use NLP to work on sentiment analysis, which consists of analyzing the meaning behind sentences. This is useful in order to understand the sentiment of clients with regard to a product or a service, based on customer reviews.

> ### NOTE
>
> All the code present in this chapter can be found at:
> https://packt.live/2yn253K.

RECURRENT NEURAL NETWORKS

Just as humans do not reset their thinking every second, neural networks that aim to understand human language should not do so either. This means that in order to understand each word from a paragraph or even a whole book, you or the model need to understand the previous words, which can help give context to words that may have different meanings.

Traditional neural networks, as we have discussed so far, are not capable of performing such tasks – hence the creation of the concept and network architecture of RNNs. As we briefly explained previously, these network architectures contain loops among the different nodes. This allows information to remain in the model for longer periods of time. Due to this, the output from the model becomes both a prediction and a memory, which will be used when the next bit of sequenced text is passed through the model.

This concept goes back to the 1980s, although it has only become popular recently thanks to advances in technology that have led to an increase in the computational power of machines and have allowed for the recollection of data, as well as the development of the concept of LSTM RNNs in the 1990s, which increased their applications. RNNs are one of the most promising network architectures out there thanks to their ability to store internal memory, which allows them to efficiently handle sequences of data and solve a wide variety of data problems.

APPLICATIONS OF RNNS

While we have made it very clear that RNNs work best with sequences of data, such as text, audio clips, and videos, it is still necessary to explain the different applications of RNNs for real-life problems.

Here are some brief explanations of the different tasks that can be performed through the use of RNNs:

- **NLP**: This refers to the ability of machines to represent human language. This is perhaps one of the most explored areas of deep learning and undoubtedly the preferred data problem when making use of RNNs. The idea is to train the network using text as input data, such as poems and books, among others, with the objective of creating a model that is capable of generating such texts.

  NLP is commonly used for the creation of chatbots (virtual assistants). By learning from previous human conversations, NLP models are able to help a person solve frequently asked questions or queries. According to this, their ability to formulate sentences is limited to what they have learned in the training process, which means that they can only answer what they have been taught.

You will probably have experienced this when trying to contact a bank through an online chat system, where, typically, you are transferred to a human operator the minute the query falls outside the conventional. Another common example of chatbots in real life are restaurants that take queries through Facebook Messenger:

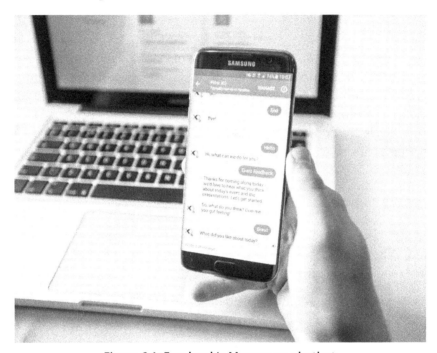

Figure 6.1: Facebook's Messenger chatbot

- **Speech recognition**: Similar to NLP, speech recognition attempts to understand and represent human language. However, the difference here is that the former (NLP) is trained on and produces the output in the form of text, while the latter (speech recognition) uses audio clips. With the proliferation of developments in this field and the interest of big companies, these models are capable of understanding different languages and even different accents and pronunciation.

A popular example of a speech recognition device is Alexa – the voice-activated virtual assistant model from Amazon:

Figure 6.2: Amazon's Alexa

- **Machine translation**: This refers to a machine's ability to translate human languages effectively. According to this, the input is the source language (for instance, Spanish) and the output is the target language (for instance, English). The main difference between NLP and machine translation is that, in the latter, the output is built after the entire input has been fed to the model.

With the rise of globalization and the popularity of leisure traveling nowadays, people require access to more than one language. Due to this, a proliferation of devices that are capable of translating between different languages has emerged. One of the latest creations in this field is Google's Pixel Buds, which can perform translations in real time:

Figure 6.3: Google's Pixel Buds

- **Time-series forecasting**: A less popular application of an RNN is the prediction of a sequence of data points in the future based on historical data. RNNs are particularly good at this task due to their ability to retain an internal memory, which allows time-series analysis to consider the different timesteps in the past to perform a prediction or a series of predictions in the future.

This is often used to foresee future income or demand, which helps a company be prepared for different scenarios. The following plot shows the forecasting of monthly sales:

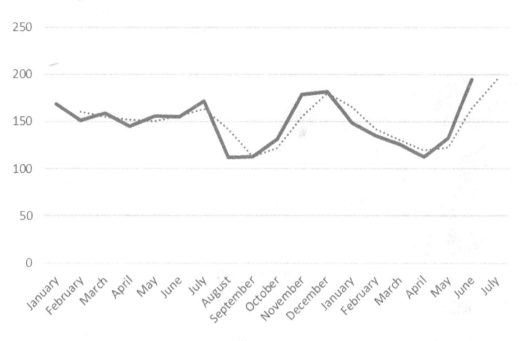

Figure 6.4: The forecasting of monthly sales (quantity)

For example, if, by forecasting the demand for several health care products, it is determined that there will be an increase in demand for one of the products and a decrease in demand for another, the company may decide to produce more of that product and less of the other.

- **Image recognition**: Coupled with CNNs, RNNs can give an image a caption or a description. This combination of models allows you to detect all the objects in the image, which determines what the image is principally made of. The output can either be a set of tags of the objects present in the image, a description of the image, or a caption of the relevant objects in the image, as shown in the following diagram:

TAGS
Dog. Plants. Grass. Flowers.

DESCRIPTION
The profile of a dog
sitting in a garden.

CAPTION
A dog in the yard.

Figure 6.5: Image recognition using RNNs

HOW DO RNNS WORK?

Simply put, RNNs take an input (x) and return an output (y). Here, the output is not only influenced by the input but also by the entire history of inputs that were fed in the past. This history of inputs is often referred to as the model's internal state or memory, which are sequences of data that follow an order and are related to one another – such as a time series, which is a sequence of data points (for example, sales) that are listed in an order (by month, for instance).

> **NOTE**
>
> Bear in mind that the general structure of an RNN may vary, depending on the problem at hand. For instance, they can be of the one-to-many type or the many-to-one type, as we mentioned in *Chapter 2*, *Building Blocks of Neural Networks*.

To understand the concept of RNNs, it is important to know the difference between RNNs and traditional neural networks. Traditional neural networks are often referred to as feedforward neural networks because the information only moves in one direction (that is, from the input to the output), without going through a node twice to perform a prediction. These networks do not have any memory of what has been fed in the past, which is why they are no good at predicting what comes next in a sequence.

On the other hand, in RNNs, information cycles through loops so that every prediction is made by considering both the input and the memory from previous predictions. It works by copying the output of each prediction and passing it back into the network for the subsequent prediction. In this way, RNNs have two inputs – the present value and past information:

A **B**

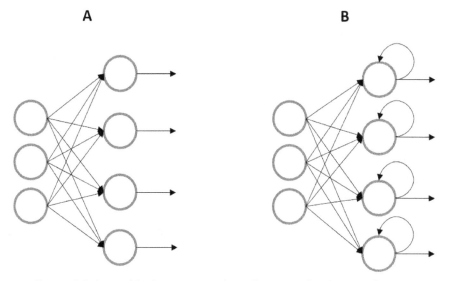

Figure 6.6: A graphical representation of a network, where A shows a
feedforward neural network and B shows an RNN

NOTE

The internal memory of traditional RNNs is short-term only. However, we will explore an architecture that is capable of storing long-term and short-term memory later.

By using information from previous predictions, the network is trained with a sequence of ordered data that allows it to predict the following step. This is achieved by combining the current information and the output from the previous step into a single operation. This can be seen in the following diagram. The output from this operation will become the prediction, as well as part of the input for the subsequent prediction:

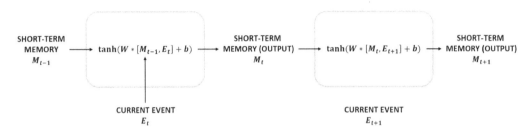

Figure 6.7: An RNN computation for each prediction

As you can see, the operation that occurs inside a node is that of any other neural network; initially, the data is passed through a linear function. The weights and biases are the parameters to be updated during the training process. Next, the linearity of this output is broken using an activation function. In this case, this is the *tanh* function, as several studies have shown that it achieves better results for most sequenced data problems:

$$M(output)_t = tanh\left(W * [M_{t-1}, E_t] + b\right)$$

Figure 6.8: A mathematical computation of traditional RNNs

Here, M_{t-1} refers to the memory that is derived from the previous prediction, W and b are the weights and biases, and E_t refers to the current event.

With this learning process in mind, let's consider the sales data of a product from the last 2 years. RNNs are capable of predicting the next month's sales because, by storing the information from the last couple of months, they are able to check whether sales have been increasing or decreasing.

Using *Figure 6.7*, the prediction of the next month could be handled by taking the last month's sales (that is, the current event) and the short-term memory (which is a representation of the data from the last couple of months) and combining them. The output from this operation will contain both the prediction of the next month and some relevant information from the last couple of months, which will, in turn, become the new short-term memory for the subsequent prediction.

Moreover, it is important to mention that some RNN architectures, such as the LSTM network, will also be able to consider data from 2 years ago or even much earlier (since it stores long-term memory). This will let the network know whether a decrease during a particular month is likely to continue to decrease or start to increase. We will explore this topic in more detail later on.

INPUT AND TARGETS FOR SEQUENCED DATA

Considering that the objective is to predict the following element in a sequence, the target matrix is typically the same information as the input data, with the target being one step ahead.

This means that the input variable should contain all the data points of the sequence, except for the last value, while the target variable should contain all the data points of the sequence, except for the first value – that is, the first value of the target variable should be the second one of the input variable, and so on, as shown in the following diagram:

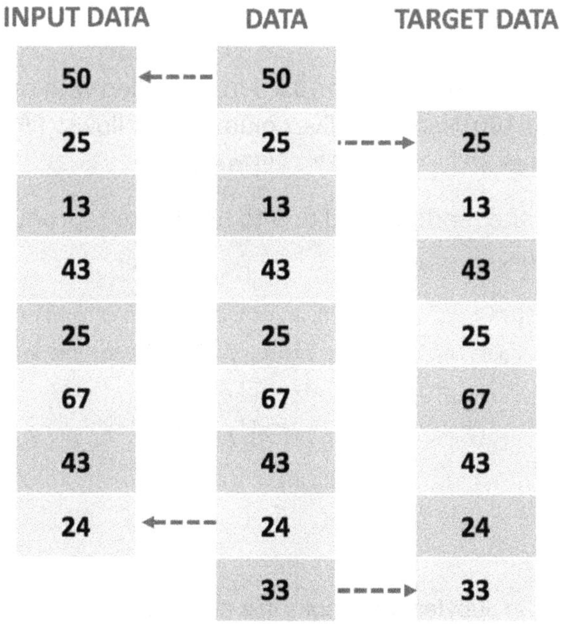

INPUT DATA	DATA	TARGET DATA
50 ←----	50	
25	25 ----→	25
13	13	13
43	43	43
25	25	25
67	67	67
43	43	43
24 ←----	24	24
	33 ----→	33

Figure 6.9: Input and target variables for a sequenced data problem

EXERCISE 6.01: CREATING THE INPUT AND TARGET VARIABLES FOR A SEQUENCED DATA PROBLEM

In this exercise, using a dummy dataset, you will learn how to create the inputs and target variables that can be used to solve a sequenced data problem. Follow these steps to complete this exercise:

> **NOTE**
>
> For the exercises and activities in this chapter, you will need to have Python 3.7, Jupyter 6.0, Matplotlib 3.1, NumPy 1.17, Pandas 0.25, and PyTorch 1.3+ (preferably PyTorch 1.4) installed on your local machine.

1. Import the following libraries:

```
import pandas as pd
import numpy as np
import torch
```

2. Create a Pandas DataFrame that's 10 x 5 in size, filled with random numbers ranging from 0 to 100. Name the five columns as follows: **["Week1", "Week2", "Week3", "Week4", "Week5"].**

   Make sure to set the random seed to **0** to be able to reproduce the results shown in this book:

```
np.random.seed(0)
data = pd.DataFrame(np.random.randint(0,100,size=(10, 5)),
                    columns=['Week1','Week2','Week3',\
                             'Week4','Week5'])
data
```

> **NOTE**
>
> As a reminder, in Jupyter Notebooks it is possible to print the value of a variable without the need of the print function. In other programming platforms, you may be required to use the print function.

The resulting DataFrame is as follows:

	Week1	Week2	Week3	Week4	Week5
0	44	47	64	67	67
1	9	83	21	36	87
2	70	88	88	12	58
3	65	39	87	46	88
4	81	37	25	77	72
5	9	20	80	69	79
6	47	64	82	99	88
7	49	29	19	19	14
8	39	32	65	9	57
9	32	31	74	23	35

Figure 6.10: The created DataFrame

3. Create an input and a target variable, considering that the input variable should contain all the values of all the instances, except the last column of data. The target variable should contain all the values of all the instances, except the first column:

```
inputs = data.iloc[:,:-1]
targets = inputs.shift(-1, axis="columns", \
                 fill_value=data.iloc[:,-1:])
```

4. Print the input variable to verify its contents, as follows:

```
inputs
```

The inputs variable should look as follows:

	Week1	Week2	Week3	Week4
0	47	64	67	67
1	83	21	36	87
2	88	88	12	58
3	39	87	46	88
4	37	25	77	72
5	20	80	69	79
6	64	82	99	88
7	29	19	19	14
8	32	65	9	57
9	31	74	23	35

Figure 6.11: Input variable

5. Print the resulting target variable using the following code:

```
targets
```

Running the preceding code will display the following output:

	Week1	Week2	Week3	Week4
0	47	64	67	67
1	83	21	36	87
2	88	88	12	58
3	39	87	46	88
4	37	25	77	72
5	20	80	69	79
6	64	82	99	88
7	29	19	19	14
8	32	65	9	57
9	31	74	23	35

Figure 6.12: Target variable

NOTE

To access the source code for this specific section, please refer to https://packt.live/2VQ5OjB.

You can also run this example online at https://packt.live/31yCotG. You must execute the entire Notebook in order to get the desired result.

RNNS IN PYTORCH

In PyTorch, similar to any other layer, the recurrent layers are defined in a single line of code. This will then be called inside the forward function of the network, as shown in the following code:

```
class RNN(nn.Module):
    def __init__(self, input_size, hidden_size, num_layers):
        super().__init__()
        self.hidden_size = hidden_size
        self.rnn = nn.RNN(input_size, hidden_size, num_layers,\
                          batch_first=True)
        self.output = nn.Linear(hidden_size, 1)

    def forward(self, x, hidden):
        out, hidden = self.rnn(x, hidden)
        out = out.view(-1, self.hidden_size)
        out = self.output(out)
        return out, hidden
```

Here, the recurrent layer must be defined as taking arguments for the number of expected features in the input (**input_size**); the number of features in the hidden state, which is defined by the user (**hidden_size**); and the number of recurrent layers (**num_layers**).

> **NOTE**
>
> In a similar way to any other neural network, the hidden size refers to the number of nodes (neurons) in that layer.

The **batch_first** argument is set to **True** to define that the input and output tensors are in the form of batches, sequences, and features.

In the **forward** function, the input is passed through the recurrent layers and the output from those layers is flattened out so that it can be passed through the fully connected layer. It is worth mentioning that the information is passed through the RNN layers, along with a hidden state (memory).

Moreover, the training of such a network can be handled as follows:

```
for i in range(1, epochs+1):

    hidden = None

    for inputs, targets in batches:
        pred, hidden = model(inputs, hidden)

        loss = loss_function(pred, targets)
        optimizer.zero_grad()
        loss.backward()
        optimizer.step()
```

For each epoch, the hidden state is initialized to **none**. This is because, in each epoch, the network will try to map the inputs to the targets (when given a set of parameters). This mapping should occur without any bias (hidden state) from the previous runs through the dataset.

Next, a **for** loop is used to go through the different batches of data. Inside this loop, a prediction is made and a hidden state is saved, which will be used as the input for the following batch.

Finally, the loss function is calculated, which is used to update the parameters of the network. Then, the process starts again until the desired number of epochs has been reached.

ACTIVITY 6.01: USING A SIMPLE RNN FOR A TIME SERIES PREDICTION

For this activity, you will use a simple RNN to solve a time series problem. Let's consider the following scenario: your company wants to be able to predict the demand, ahead of time, for all its products. This is because it takes quite some time to produce each product and the procedure costs a lot of money. Consequently, they do not wish to spend money and time in production unless the product is likely to be sold. In order to predict the future demand, they have provided you with a dataset containing the weekly demand (in sales transactions) for all the products from last year's sales. Follow these steps to complete this activity:

> **NOTE**
>
> The CSV file containing the dataset that will be used in this activity can be found at https://packt.live/2K5pQQK. It is also available online at https://archive.ics.uci.edu/ml/datasets/Sales_Transactions_Dataset_Weekly.
>
> The dataset and related analysis was first published here: Tan S.C., Lau J.P.S. (2014) Time Series Clustering: A Superior Alternative for Market Basket Analysis. In: Herawan T., Deris M., Abawajy J. (eds) Proceedings of the First International Conference on Advanced Data and Information Engineering (DaEng-2013). Lecture Notes in Electrical Engineering, vol 285. Springer, Singapore.

1. Import the required libraries.

2. Load the dataset and slice it so that it contains all the rows but only the columns from index 1 to 52.

3. Plot the weekly sales transactions of five randomly chosen products from the entire dataset. Use a random seed of **0** when doing random sampling in order to achieve the same results as in the current activity.

4. Create the **inputs** and **targets** variables, which will be fed to the network to create the model. These variables should be of the same shape and be converted into PyTorch tensors.

    The **inputs** variable should contain the data for all the products for all the weeks, except the last week because the idea of the model is to predict this final week.

    The **targets** variable should be one step ahead of the **inputs** variable; that is, the first value of the **targets** variable should be the second one of the inputs variable, and so on, until the last value of the **targets** variable (which should be the last week that was left outside of the **inputs** variable).

5. Create a class containing the architecture of the network; note that the output size of the fully connected layer should be 1.

6. Instantiate the class function containing the model. Feed the input size, the number of neurons in each recurrent layer (10), and the number of recurrent layers (1).

7. Define a loss function, an optimization algorithm, and the number of epochs to train the network. Use the Mean Squared Error loss function, the Adam optimizer, and 10,000 epochs for this.

8. Use a **for** loop to perform the training process by going through all the epochs. In each epoch, a prediction must be made, along with the subsequent calculation of the loss function and the optimization of the parameters of the network. Then, save the loss of each of the epochs.

9. Plot the losses of all the epochs.

10. Using a scatter plot, display the predictions that were obtained in the last epoch of the training process against the ground truth values (that is, the sales transactions of the last week).

> **NOTE**
>
> The solution to this activity can be found on page 284.

LONG SHORT-TERM MEMORY NETWORKS

As we mentioned previously, RNNs store short-term memory only. This is an issue when dealing with long sequences of data, where the network will have trouble carrying the information from the earlier steps to the final ones.

For instance, take the poem "The Raven," which was written by the famous poet Edgar Alan Poe and is over 1,000 words long. Attempting to process it using a traditional RNN, with the objective of creating a similar related poem, will result in the model leaving out crucial information from the first couple of paragraphs. This, in turn, may result in an output that is unrelated to the initial subject of the poem. For instance, it could ignore that the event occurred at night, and so make the new poem not very scary.

This inability to hold long-term memory occurs because traditional RNNs suffer from a problem called vanishing gradients. This occurs when the gradients, which are used to update the parameters of the network to minimize the loss function, become extremely small so that they no longer contribute to the learning process of the network. This typically occurs in the first few layers of the networks, making the network forget what it saw a while ago.

Because of this, **LSTM** networks were developed. LSTM networks can remember information over long periods of time as they store their internal memory in a similar way to a computer; that is, they read, write, and delete information as needed, which is achieved through the use of gates.

These gates help the network decide what information to keep and what information to delete from the memory (whether to open the gate) based on the importance that it assigns to each bit of information. This is extremely useful because it not only allows for more information to be stored (as long-term memory), but it also helps throw away useless information that may alter the result of a prediction, such as the articles in a sentence.

APPLICATIONS OF LSTM NETWORKS

Besides the applications we explained previously, the ability of LSTM networks to store long-term information has allowed data scientists to work on complex data problems that make use of large sequences of data as inputs, some of which will be explained here:

- **Text generation**: Generating any text, such as the text you are reading here, can be converted into the task of an LSTM network. This works by selecting each letter based on all the previous letters. Networks that perform this task are trained with large texts, such as those of famous books. This is because the final model will create a text that is similar to the style of writing of the one that it was trained on. For instance, a model that is trained on a poem will have a narrative that is different from the one you would expect in a conversation with a neighbor.

- **Music generation**: Just as a sequence of text can be inputted into the network with the objective of generating similar new text, a sequence of notes can also be fed into the network to generate new sequences of musical notes. Keeping track of the previous notes will help achieve a harmonized melody, rather than just a series of random musical notes. For example, feeding an audio file with a popular song from The Beatles will result in a sequence of musical notes that resembles the harmony of the group.

- **Handwriting generation and recognition**: Here, each letter is also a product of all the previous letters, which, in turn, will result in a set of handwritten letters that have a meaning. Likewise, LSTM networks can also be used to recognize handwritten texts, where the prediction of one letter will depend on all the previously predicted letters. For instance, recognizing an ugly-looking handwritten letter can be easier when considering the previous letters, as well as the entire paragraph, as it helps narrow down the prediction according to the context.

HOW DO LSTM NETWORKS WORK?

So far, it has been made clear that what differentiates LSTM networks from traditional RNNs is their ability to have a long-term memory. However, it is important to mention that, as time passes, very old information is less likely to influence the next output. Considering this, LSTM networks also have the ability to take into account the distance between data bits and the underlying context in order to also make the decision to forget some piece of information that is no longer relevant.

So, how do LSTM networks decide when to remember and when to forget? Different from traditional RNNs, where only one calculation is performed in each node, LSTM networks perform four different calculations that allow for interaction between the different inputs of the network (that is, the current event, the short-term memory, and the long-term memory) to arrive at an outcome.

To understand the process behind LSTM networks, consider the four gates that are used to manage the information in the network, as shown in the following diagram:

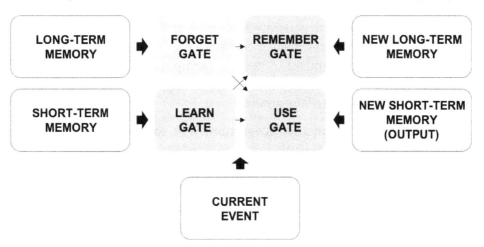

Figure 6.13: LSTM network gates

The functionality of each of the gates in the preceding diagram can be explained as follows:

- **Learn gate**: Both the short-term memory (also known as the hidden state) and the current event go into the learn gate, where the information is analyzed, and any unwanted information is ignored. Mathematically, this happens by combining both the short-term memory and the current event using a linear function and an activation function (*tanh*). The output from this is multiplied by an ignore factor, which removes any irrelevant information. To calculate the ignore factor, the short-term memory and the current event are passed through a linear function. Then, they are squeezed together by the **sigmoid** activation function:

$$M_t = tanh(W * [STM_{t-1}, E_t] + b)$$

$$I_t = \sigma(W * [STM_{t-1}, E_t] + b)$$

$$L_t = M_t * I_t$$

Figure 6.14: The mathematical computations that occur in the learn gate

Here, STM_{t-1} refers to the short-term memory that is derived from the previous prediction, W and b are the weights and biases, and E_t refers to the current event.

- **Forget gate**: The long-term memory (also known as the cell state) goes into the forget gate, where some information is removed. This is achieved by multiplying the long-term memory and the forget factor. To calculate the forget factor, the short-term memory and the current event are passed through a linear function and an activation function (**sigmoid**):

$$FF_t = \sigma(W * [STM_{t-1}, E_t] + b)$$

$$F_t = LTM_{t-1} * FF_t$$

Figure 6.15: The mathematical computations that occur in the forget gate

Here, STM_{t-1} refers to the short-term memory that is derived from the previous prediction, LTM_{t-1} is the long-term memory that is derived from the previous prediction, W and b are the weights and biases, and E_t refers to the current event.

- **Remember gate**: The long-term memory that was not forgotten in the forget gate and the information that was kept from the learn gate are joined together in the remember gate, which will become the new long-term memory. Mathematically, this is achieved by summing the output from the learn and forget gates:

$$R_t = L_t + F_t$$

Figure 6.16: The mathematical computation that occurs in the remember gate

Here, L_t refers to the output from the learn gate, while F_t refers to the output from the forget gate.

- **Use gate**: This is also known as the output gate. Here, the information from both the learn and forget gates are joined together in the use gate. This gate makes use of all the relevant information to perform a prediction, which also becomes the new short-term memory.

This is achieved in three steps. First, it applies a linear and an activation function (*tanh*) to the output from the forget gate. Second, it applies a linear and an activation function (*sigmoid*) to both the short-term memory and the current event. Third, it multiplies the output from the previous steps. The output from the third step will be the new short-term memory and the prediction from the current step:

$$UF_t = tanh(W * F_t + b)$$

$$US_t = \sigma(W * [STM_{t-1}, E_t] + b)$$

$$U_t = UF_t * US_t$$

Figure 6.17: The mathematical computations that occur in the use gate

Here, STM_{t-1} refers to the short-term memory that is derived from the previous prediction, W and b are the weights and biases, and E_t refers to the current event.

> **NOTE**
>
> Although the use of the different activation functions and mathematical operators seems arbitrary, it is done this way because it has been proved to work on most data problems that deal with large sequences of data.

The preceding process is done for every single prediction that is performed by the model. For instance, for a model built to create literary pieces, the process of learning, forgetting, remembering, and using the information is performed for every single letter that will be produced by the model, as shown in the following diagram:

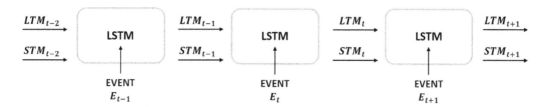

Figure 6.18: An LSTM network process through time

LSTM NETWORKS IN PYTORCH

The process of defining the LSTM network architecture in PyTorch is similar to that of any other neural network that we have discussed so far. However, it is important to note that, when dealing with sequences of data that are different from those of numbers, there is some preprocessing required in order to feed the network with data that it can understand and process.

Considering this, we need to explain the general steps for training a model to be able to take text data as inputs and retrieve a new piece of textual data. It is important to mention that not all the steps explained here are strictly required, but as a group, they make clean and reusable code for using LSTMs with textual data.

PREPROCESSING THE INPUT DATA

The first step is to load the text file into the code. This data will go through a series of transformations in order to be fed into the model properly. This is necessary because neural networks perform a series of mathematical computations to arrive at an output, which means that all the inputs must be numerical. Additionally, it is also good practice to feed the data in batches to the model, rather than all at once, as it helps reduce the training times, especially for long datasets. These transformations will be explained in the following subsections.

NUMBERED LABELS

First, a list of unduplicated characters is obtained from the input data. Each of these characters is assigned a number. Then, the input data is encoded by replacing each character with the assigned number, considering that the same letter must always be represented by the same number. For instance, the word "hello" will be encoded into 12334, given the following mapping of characters and numbers:

$$mapping = \{"h":1,"e":2,"l":3,"o":4\}$$

Figure 6.19: The mapping of characters and numbers

The preceding output can be achieved through the following code snippet:

```
text = "this is a test text!"
chars = list(set(text))
indexer = {char: index for (index, char) \
           in enumerate(chars)}

indexed_data = []
for c in text:
    indexed_data.append(indexer[c])
```

The second line of code creates a list containing the alphabet of the text (that is, the letters and characters in the text sequence). Next, a dictionary is created using each letter or character as the key and the numerical representation associated with it as the value. Finally, by performing a **for** loop through the text, it is possible to replace each letter or character with its numerical representation, thereby converting the text into a numeric matrix.

GENERATING THE BATCHES

For RNNs, batches are created using two variables: the number of sequences per batch and the length of each sequence. These values are used to divide the data into matrices, which will help speed up the calculations.

Using a dataset of 24 integers, with the number of sequences per batch set to 2 and the sequence length equal to 4, the division works as follows:

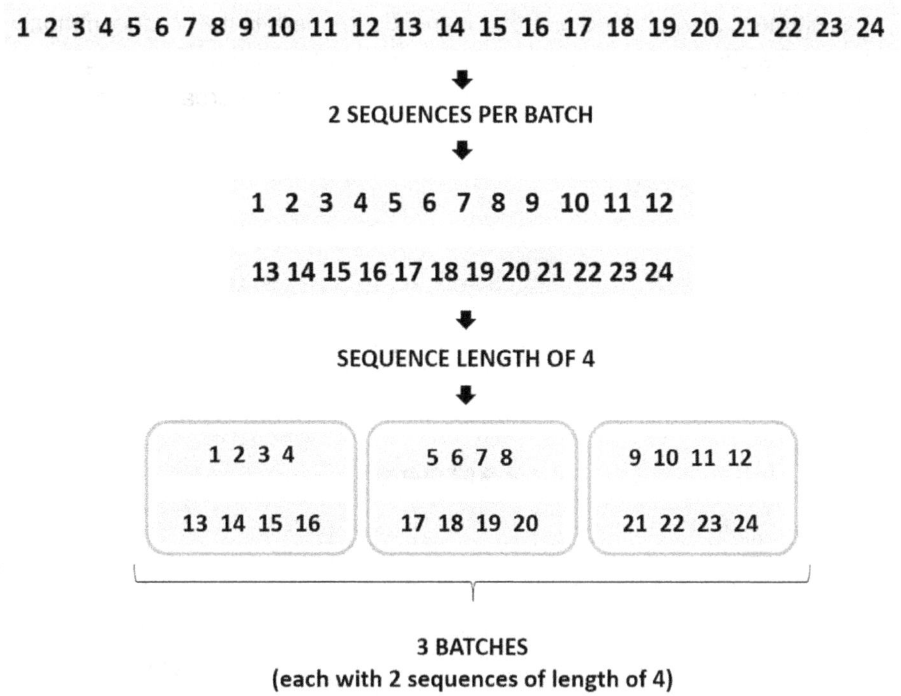

1 2 3 4 5 6 7 8 9 10 11 12 13 14 15 16 17 18 19 20 21 22 23 24

⬇

2 SEQUENCES PER BATCH

⬇

1 2 3 4 5 6 7 8 9 10 11 12

13 14 15 16 17 18 19 20 21 22 23 24

⬇

SEQUENCE LENGTH OF 4

⬇

| 1 2 3 4 | 5 6 7 8 | 9 10 11 12 |
| 13 14 15 16 | 17 18 19 20 | 21 22 23 24 |

3 BATCHES
(each with 2 sequences of length of 4)

Figure 6.20: Batch generation for RNNs

As shown in the preceding diagram, three batches are created – each of these batches containing two sequences with a length of 4.

This batch generation process should be done for **x** and for **y**, where the former is the input to the network and the latter represents the targets. According to this, the idea of the network is to find a way to map the relationship between **x** and **y**, considering that **y** will be one step ahead of **x**.

The batches for **x** are created by following the methodology explained in the previous diagram. Then, the batches of **y** will be created so that they are the same length as the ones for **x**. This is because the first element of **y** will be the second element of **x** and so on, until the last element of **y** (which will be the first element of **x**):

> **NOTE**
>
> There are a number of different approaches that you can use to fill in the last element of **y**, and the one that is mentioned here is the most commonly used. The choice of approach is often a matter of preference, although some data problems may benefit more from a certain approach than from others.

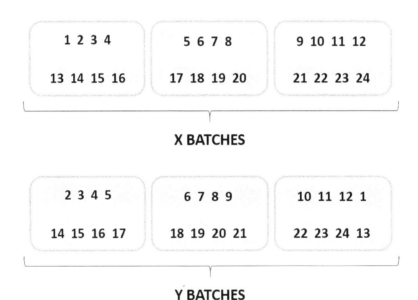

Figure 6.21: A representation of the batches for x and y

The generation of batches can be achieved through the following code snippet:

```
x = np.array(indexed_data).reshape((2,-1))
for b in range(0, x.shape[1], 5):
    batch = x[:,b:b+5]
    print(batch)
```

First, the numeric matrix is divided into sequences (as many as you desire). Next, using a **for** loop, it is possible to partition the sequenced data into batches of a specified length. By printing the **batch** variable, it is possible to observe the result.

> **NOTE**
>
> Although generating batches is considered part of preprocessing the data, it is often programmed inside the **for** loop of the training process.

ONE-HOT ENCODING

Converting all the characters into numbers is not enough to feed them into the model. This is because this approximation introduces some bias to your model since the characters that are converted into higher numerical values will be evaluated as more important. To avoid this, it is good practice to encode the different batches as one-hot matrices. This consists of creating a three-dimensional matrix with zeros and ones, where zero represents the absence of an event and one refers to the presence of an event. The final shape of the matrix should be *one hot = [number of sequences, sequence length, number of characters]*.

This means that, for every element in the batch, it will create a sequence of values of a length equal to the total number of characters in the entire text. For every character, it will place a zero, except for the one that is present in that position (where it will place a one).

> **NOTE**
>
> You can find out more about one-hot encoding at
> https://www.geeksforgeeks.org/ml-one-hot-encoding-of-datasets-in-python/.

This can be achieved through the following code snippet:

```
batch = np.array([[2 4 7 6 5]
                  [2 1 6 2 5]])

batch_flatten = batch.flatten()
onehot_flat = np.zeros((batch.shape[0] \
                        * batch.shape[1],len(indexer)))
onehot_flat[range(len(batch_flatten)), batch_flatten] = 1
onehot = onehot_flat.reshape((batch.shape[0],\
                             batch.shape[1], -1))
```

First, the two-dimensional batch is flattened. Next, a matrix is created and filled with zeros. Zeros are replaced by ones when we need to represent the correct character in a given position. Finally, the flattened dimension is expanded again.

EXERCISE 6.02: PREPROCESSING THE INPUT DATA AND CREATING A ONE-HOT MATRIX

In this exercise, you will preprocess a text snippet, which will then be converted into a one-hot matrix. Follow these steps to complete this exercise:

1. Import NumPy:

    ```
    import numpy as np
    ```

2. Create a variable named **text**, which will contain the text sample **"Hello World!"**:

    ```
    text = "Hello World!"
    ```

3. Create a dictionary by mapping each letter to a number:

    ```
    chars = list(set(text))
    indexer = {char: index for (index, char) \
               in enumerate(chars)}
    print(indexer)
    ```

Running the preceding code will result in the following output:

```
{'d': 0, 'o': 1, 'H': 2, ' ': 3, 'e': 4, 'W': 5, '!': 6, 'l': 7, 'r':
8}
```

4. Encode your text sample with the numbers we defined in the previous step:

```
encoded = []
for c in text:
    encoded.append(indexer[c])
```

5. Convert the encoded variable into a NumPy array and reshape it so that the sentence is divided into two sequences of the same size:

```
encoded = np.array(encoded).reshape(2,-1)
encoded
```

Running the preceding code will result in the following output:

```
array([[2, 4, 7, 7, 1, 3],
       [5, 1, 8, 7, 0, 6]])
```

6. Define a function that takes an array of numbers and creates a one-hot matrix:

```
def index2onehot(batch):

    batch_flatten = batch.flatten()
    onehot_flat = np.zeros((batch.shape[0] \
                           * batch.shape[1], len(indexer)))
    onehot_flat[range(len(batch_flatten)), \
               batch_flatten] = 1
    onehot = onehot_flat.reshape((batch.shape[0], \
                           batch.shape[1], -1))

    return onehot
```

7. Convert the encoded array into a one-hot matrix by passing it through the previously defined function:

```
one_hot = index2onehot(encoded)
one_hot
```

The output should appear as follows:

```
array([[[0., 0., 1., 0., 0., 0., 0., 0., 0.],
        [0., 0., 0., 0., 1., 0., 0., 0., 0.],
        [0., 0., 0., 0., 0., 0., 0., 1., 0.],
        [0., 0., 0., 0., 0., 0., 0., 1., 0.],
        [0., 1., 0., 0., 0., 0., 0., 0., 0.],
        [0., 0., 0., 1., 0., 0., 0., 0., 0.]],

       [[0., 0., 0., 0., 0., 1., 0., 0., 0.],
        [0., 1., 0., 0., 0., 0., 0., 0., 0.],
        [0., 0., 0., 0., 0., 0., 0., 0., 1.],
        [0., 0., 0., 0., 0., 0., 0., 1., 0.],
        [1., 0., 0., 0., 0., 0., 0., 0., 0.],
        [0., 0., 0., 0., 0., 0., 1., 0., 0.]]])
```

Figure 6.22: One-hot representation of sample text

NOTE

To access the source code for this specific section, please refer to https://packt.live/2ZpGvFJ.

You can also run this example online at https://packt.live/38foCxD. You must execute the entire Notebook in order to get the desired result.

You have successfully converted some sample text into a one-hot matrix.

BUILDING THE ARCHITECTURE

Similar to other neural networks, an LSTM layer is easily defined in a single line of code. Nevertheless, the class containing the architecture of the network must include a function that allows the hidden and cell state (that is, both memories of the network) to be initialized. An example of an LSTM network architecture is as follows:

```python
class LSTM(nn.Module):
    def __init__(self, char_length, hidden_size, n_layers):
        super().__init__()
        self.hidden_size = hidden_size
        self.n_layers = n_layers
        self.lstm = nn.LSTM(char_length, hidden_size, \
                            n_layers, batch_first=True)
        self.output = nn.Linear(hidden_size, char_length)

    def forward(self, x, states):
        out, states = self.lstm(x, states)
        out = out.contiguous().view(-1, self.hidden_size)
        out = self.output(out)

        return out, states

    def init_states(self, batch_size):
        hidden = next(self.parameters()).data.new(self.n_layers, \
                    batch_size, self.hidden_size).zero_()
        cell = next(self.parameters()).data.new(self.n_layers, \
                    batch_size, self.hidden_size).zero_()
        states = (hidden, cell)

        return states
```

> **NOTE**
>
> Again, the **batch_first** argument is set to **True** when the input and output tensors are in the form of batches, sequences, and features. Otherwise, there is no need to define it as its default value is **False**.

As we can see, the LSTM layers are defined in a single line taking the following as arguments:

- The number of features in the input data (that is, the number of non-duplicated characters)

- The number of hidden dimensions (neurons)

- The number of LSTM layers

The **forward** function, as with any other network, defines the way that data is moved through the layers in a **forward** pass.

Finally, a function is defined to initialize the hidden and cell states to zero in every epoch. This is achieved by **next(self.parameters()).data.new()**, which grabs the first parameter of the model and creates a new tensor of the same type with the specified dimensions inside the parentheses, which is then filled with zeros. Both the **hidden** and **cell** states are fed into the model as a tuple.

TRAINING THE MODEL

Once the loss function and the optimization algorithm have been defined, it is time to train the model. This is achieved by following a very similar approach to the one that is used for other neural network architectures, as shown in the following code snippet:

```
# Step 1: for through epochs
for e in range(1, epochs+1):

    # Step 2: Memory initialized
    states = model.init_states(n_seq)

    # Step 3: for loop to split data in batches.
    for b in range(0, x.shape[1], seq_length):
        x_batch = x[:,b:b+seq_length]

        if b == x.shape[1] - seq_length:
            y_batch = x[:,b+1:b+seq_length]
            y_batch = np.hstack((y_batch, indexer["."] \
```

```
                        * np.ones((y_batch.shape[0],1))))
    else:
        y_batch = x[:,b+1:b+seq_length+1]


    """
    Step 4: input data is converted to one-hot matrix.
    Inputs and targets are converted to tensors.
    """
    x_onehot = torch.Tensor(index2onehot(x_batch))
    y = torch.Tensor(y_batch).view(n_seq * seq_length)


    """
    Step 5: get a prediction and perform the
    backward propagation
    """
    pred, states = model(x_onehot, states)
    loss = loss_function(pred, y.long())
    optimizer.zero_grad()
    loss.backward(retain_graph=True)
    optimizer.step()
```

As shown in the preceding code, the steps are as follows:

1. It is necessary to go through the data several times in order to arrive at a better model—hence, the necessity to set a number of epochs.

2. In each epoch, the hidden and cell states must be initialized. This is achieved by making a call to the function that was previously created in the class.

3. Data is fed into the model in batches using a **for** loop. An **if** statement is used to determine whether it is the last batch in order to add a dot to the end of a sentence, which will be used to represent a period.

4. The input data is converted into a one-hot matrix. Both the input and target are converted into PyTorch tensors.

5. The output from the network is obtained by calling the model over a batch of data. Then, the loss function is calculated, and the parameters are optimized.

PERFORMING PREDICTIONS

It is good practice to provide the first couple of characters to the trained model in order to perform a prediction that has some sort of purpose (for instance, a paragraph starting with the words "once upon a time"). These initial characters should be fed to the model without performing any prediction, but with the purpose of generating a memory. Next, the previous character and the memory are fed into the network and the output is passed through a **softmax** function in order to calculate the probability of each character being the next character in the sequence. Finally, from the characters with higher probabilities, one is randomly chosen.

This can be achieved through the following code snippet:

```
# Step 1
starter = "This is the starter text"
states = None

# Step 2
for ch in starter:
    x = np.array([[indexer[ch]]])
    x = index2onehot(x)
    x = torch.Tensor(x)

    pred, states = model(x, states)

# Step 3
counter = 0
while starter[-1] != "." and counter < 50:
    counter += 1
    x = np.array([[indexer[starter[-1]]]])
    x = index2onehot(x)
    x = torch.Tensor(x)

    pred, states = model(x, states)
    pred = F.softmax(pred, dim=1)
    p, top = pred.topk(10)
    p = p.detach().numpy()[0]
    top = top.numpy()[0]
```

```
    index = np.random.choice(top, p=p/p.sum())

    # Step 4
    starter += chars[index]
    print(starter)
```

The following steps occur in the previous snippet:

1. The starting sentence is defined.

2. A **for** loop is used to feed each of the characters of the starting sentence into the model in order to update the memory of the model, prior to making a prediction.

3. A **while** loop is used to perform predictions of new characters, as long as the character count doesn't surpass 50, and until the new character is a period.

4. Each new character is added to the starting sentence to form the new text sequence.

ACTIVITY 6.02: TEXT GENERATION WITH LSTM NETWORKS

> **NOTE**
>
> The text data that will be used in this activity can be accessed for free on the internet, although you can also find it in this book's GitHub repository. The URL of the repository was mentioned in the introduction to this chapter.

In this activity, we will train an LSTM network using the book *Alice in Wonderland* so that we can feed a starting sentence into the model and have it complete the sentence. Consider the following scenario: you love things that make life easier and have decided to build a model that helps you complete sentences when you are writing an email. To do so, you have decided to train a network using a popular children's book. Follow these steps to complete this activity:

> **NOTE**
>
> It is important to mention that while the network in this activity is trained for enough iterations to display decent results, it is not trained and configured to achieve the best performance. You are encouraged to play with it to improve the performance.

1. Import the required libraries.

2. Open and read the text from *Alice in Wonderland* into the notebook. Print an extract of the first 50 characters and the total length of the text file.

3. Create a variable containing a list of the unduplicated characters in your dataset. Then, create a dictionary that maps each character to an integer, where the characters will be the keys and the integers will be the values.

4. Encode each letter of your dataset to their paired integer. Print the first 50 encoded characters and the total length of the encoded version of your dataset.

5. Create a function that takes in a batch and encodes it as a one-hot matrix.

6. Create a class that defines the architecture of the network. The class should contain an additional function that initializes the states of the LSTM layers.

7. Determine the number of batches to be created from your dataset, bearing in mind that each batch should contain 100 sequences and each should have a length of 50. Next, split the encoded data into 100 sequences.

8. Instantiate your model using 256 as the number of hidden units for a total of two recurrent layers.

9. Define the loss function and the optimization algorithms. Use the Adam optimizer and the cross-entropy loss. Train the network for 20 epochs.

> **NOTE**
>
> Depending on your resources, the training process will take a long time, which is why the recommendation is to run for only 20 epochs. However, an equivalent version of the code that can be run on the GPU is available in this book's GitHub repository. This will allow you to run for more epochs and achieve outstanding performance.

10. In each epoch, the data must be divided into batches with a sequence length of 50. This means that each epoch will have 100 sequences, each with a length of 50.

> **NOTE**
>
> Batches are created for the inputs and targets, where the latter is a copy of the former but one step ahead.

11. Plot the progress of the loss over time.

12. Feed the following sentence starter into the trained model and complete the sentence: "So she was considering in her own mind ".

> **NOTE**
>
> The solution to this activity can be found on page 290.

The final sentence will vary because there is a random factor when it comes to choosing each character; however, it should look something like this:

So she was considering in her own mind us on," said she whad se the sire. The preceding sentence does not have a meaning because the network was not trained for sufficient time (the loss function can be still be minimized) and that it selects each character at a time, without the long-term memory of the previously created words. Nevertheless, we can see that, after only 20 epochs, the network is already capable of forming some words that have meaning.

NATURAL LANGUAGE PROCESSING

Computers are good at analyzing standardized data, such as financial records or databases stored in tables. In fact, they are better than humans in doing so as they have the ability to analyze hundreds of variables at a time. On the other hand, humans are great at analyzing unstructured data, such as language, which is something that computers are not great at doing unless they have a set of rules at hand to help them understand it.

With this in mind, the biggest challenge for computers in regard to human language is that, even though a computer can be good at analyzing human language after being trained for a very long time on a very large dataset, they are still unable to understand the real meaning behind a sentence as they are neither intuitive nor capable of reading between the lines.

This means that while humans are able to understand that a sentence that says "He was on fire last night. What a great game!" refers to the performance of a player of some kind of sport, a computer will understand it in the literal sense—meaning that it will interpret it as someone having actually caught fire last night.

NLP is a subfield of **artificial intelligence** (**AI**) that works by enabling computers to understand human language. While it may be the case that humans will always be better at this task, the main objective of NLP is to bring computers closer to humans by making them understand human language.

The idea is to create models that focus on particular areas, such as machine translation and text summarization. This specialization of tasks helps the computer develop a model that is capable of solving real-life data problems, without having to deal with all the complexity of human language all at once.

One of these areas of human language understanding (one that is highly popular these days) is sentiment analysis.

SENTIMENT ANALYSIS

In general terms, sentiment analysis consists of understanding the sentiment behind the input text. It has grown increasingly popular considering that, with the proliferation of social media platforms, the quantity of messages and comments that a company receives each day has grown exponentially. This has made the task of manually revising and responding to each message in real-time impossible, which can be damaging to a company's image.

Sentiment analysis focuses on extracting the essential components of a sentence while ignoring the details. This helps solve two primary needs:

1. Identifying the key aspects of a product or service that customers care the most about.

2. Extracting the feelings behind each of these aspects to determine which ones are causing positive and negative reactions and dealing with them accordingly:

Figure 6.23: An example of a tweet

Taking the text from the preceding screenshot, a model that performs sentiment analysis is likely to pick up the following information:

- "AI" as the object of the tweet

- "Happy" as the feeling derived from it

- "Decade" as the timeframe of the sentiment over the object

As you can see, the concept of sentiment analysis can be key to any company that has an online presence as it will have the ability to respond surprisingly fast to those comments that require immediate attention and with a precision that is similar to that of a human.

As an example use case of sentiment analysis, some companies may choose to perform sentiment analysis on the vast number of messages that they receive daily in order to prioritize a response to those messages that contain complaints or negative feelings. This will not only help to mitigate the negative feelings from those particular customers; it will also help the company correct their mistakes rapidly and create a trusting relationship with their customers.

The process of performing NLP for sentiment analysis will be explained more in the next section. We will explain the concept of word embedding and the different steps that you can perform to develop such a model in PyTorch, which will be the objective of the final activity of this chapter.

SENTIMENT ANALYSIS IN PYTORCH

Building a model to perform sentiment analysis in PyTorch is fairly similar to what we have seen so far with RNNs. The difference is that, on this occasion, the text data will be processed word by word. The steps that are required to build such a model will be provided in this section.

PREPROCESSING THE INPUT DATA

As with any other data problem, you need to load the data into the code, bearing in mind that different methodologies are used for different data types. Besides converting the entire set of words into lowercase, the data undergoes some basic transformations that will allow you to feed the data into the network. The most common transformations are as follows:

- **Eliminating punctuation**: When processing text data word by word for NLP purposes, remove any punctuation. This is done to avoid taking the same word as two separate words because one of them is followed by a period, comma, or any other special character. Once this has been achieved, it is possible to define a list containing the vocabulary (that is, the entire set of words) of the input text.

  This can be done by making use of the **string** module's **punctuation** pre-initialized string, which provides a list of punctuation characters that can be used to identify them in the text sequence, as per the following code snippet:

  ```
  test = pd.Series(['Hey! This is example #1.', \
                    'Hey! This is example #2.', \
                    'Hey! This is example #3.'])


  for i in punctuation:
      test = test.str.replace(i,"")
  ```

- **Numbered labels**: Similar to the previously explained process of mapping characters, each word in the vocabulary is mapped to an integer, which will be used to replace the words of the input text in order to feed them into the network:

$$mapping = \{"once":1,"upon":2,"a":3,"time":4\}$$

Figure 6.24: The mapping of words and numbers

Instead of performing one-hot encoding, PyTorch allows you to embed words in a single line of code that can be defined inside the class containing the network architecture (which will be explained next).

BUILDING THE ARCHITECTURE

Again, the process of defining the network architecture is fairly similar to what we have studied so far. However, as we mentioned previously, the network should also include an embedding layer that will take the input data (that has been converted into a numeric representation) and assign a degree of relevance to each word. That is to say, the values will be updated during the training process until the most relevant words are weighted more highly.

The following is an example of an architecture:

```python
class LSTM(nn.Module):
    def __init__(self, vocab_size, embed_dim, \
                 hidden_size, n_layers):
        super().__init__()
        self.hidden_size = hidden_size
        self.embedding = nn.Embedding(vocab_size, embed_dim)
        self.lstm = nn.LSTM(embed_dim, hidden_size, n_layers)
        self.output = nn.Linear(hidden_size, 1)
    def forward(self, x, states):
        out = self.embedding(x)
        out, states = self.lstm(out, states)
        out = out.contiguous().view(-1, self.hidden_size)
        out = self.output(out)

        return out, states
```

As you can see, the embedding layer will take the length of the entire vocabulary as an argument, as well as an embedding dimension that is set by the user. This embedding dimension will be the input size of the LSTM layers. The rest of the architecture will remain the same as before.

TRAINING THE MODEL

Finally, after you have defined a loss function and an optimization algorithm, the process of training the model is the same as any other neural network. Data may be split into different sets, depending on the needs and purpose of the study. Next, you must set the number of epochs and the methodology to split the data into batches. The memory of the network is typically kept while processing each batch of data but is then initialized to zero in every epoch. The output from the network is obtained by calling the model over a batch of data. Then, the loss function is calculated, and the parameters are optimized.

ACTIVITY 6.03: PERFORMING NLP FOR SENTIMENT ANALYSIS

The dataset that you will use for this activity is called the *Sentiment Labelled Sentences Dataset* and is available at the UC Irvine Machine Learning Repository.

> **NOTE**
>
> The dataset for this activity can be found at https://packt.live/2z2LYc5. It is also available online at https://archive.ics.uci.edu/ml/datasets/Sentiment+Labelled+Sentences.
>
> The dataset and related analysis was first published here: From Group to Individual Labels using Deep Features, Kotzias et. al,. KDD 2015 [https://doi.org/10.1145/2783258.2783380]

In this activity, an LSTM network will be used to analyze a set of reviews to determine the sentiment behind them. Let's consider the following scenario: you work in the public relations department of an internet provider and the process of reviewing every inquiry that you get on the company's social media profiles is taking a long time. The biggest problem is that the customers that are having issues with the service have less patience than those who do not, so you need to prioritize your responses so that you can address them first. As you enjoy programming in your free time, you have decided to try to build a neural network that is capable of determining whether a message is negative or positive. Follow these steps to complete this activity:

> **NOTE**
>
> It is important to mention that the data in this activity is not divided into different sets of data that allow the model to be fine-tuned and tested. This is because the main focus of this activity is to implement the process of creating a model that is capable of performing sentiment analysis.

1. Import the required libraries.

2. Load the dataset containing a set of 1,000 product reviews from Amazon, which are paired with a label of 0 (for negative reviews) or 1 (for positive reviews). Separate the data into two variables – one containing the reviews and the other containing the labels.

3. Remove the punctuation from the reviews.

4. Create a variable containing the vocabulary of the entire set of reviews. Additionally, create a dictionary that maps each word to an integer, where the words will be the keys and the integers will be the values.

5. Encode the reviews data by replacing each word in a review with its paired integer.

6. Create a class containing the architecture of the network. Make sure that you include an embedding layer.

> ### NOTE
> Since the data won't be fed in batches during the training process, there is no need to return the states in the **forward** function. However, this does not mean that the model will not have a memory, but rather that the memory is used to process each review individually since one review is not dependent on the next one.

7. Instantiate the model using 64 embedding dimensions and 128 neurons for 3 LSTM layers.

8. Define the loss function, an optimization algorithm, and the number of epochs to train for. For example, you can use binary cross-entropy loss as the loss function, the Adam optimizer, and train for 10 epochs.

9. Create a **for** loop that goes through the different epochs and through every single review individually. For each review, perform a prediction, calculate the loss function, and update the parameters of the network. Additionally, calculate the accuracy of the network on that training data.

10. Plot the progress of the loss and accuracy over time.

The final plot of accuracy will look as follows:

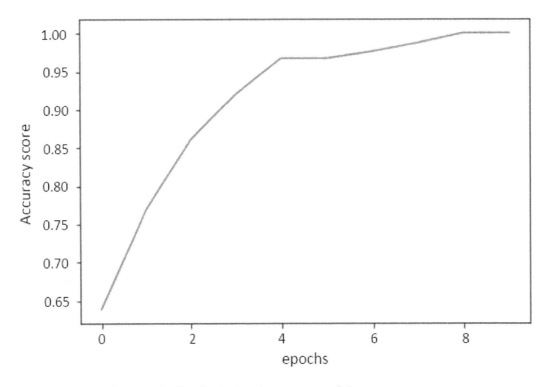

Figure 6.25: Plot displaying the progress of the accuracy score

> **NOTE**
>
> The solution to this activity can be found on page 298.

SUMMARY

In this chapter, we discussed RNNs. This type of neural network was developed in order to solve problems related to data in sequences. This means that a single instance does not contain all the relevant information since this depends on information from the previous instances.

There are several applications that fit this type of description. For example, a specific portion of text (or speech) may not mean much without the context of the rest of the text. However, even though NLP has been explored the most with RNNs, there are other applications where the context of the text is important, such as forecasting, video processing, or music-related problems.

An RNN works in a very clever way; the network not only outputs a result but also one or more values that are often referred to as memory. This memory value is used as input for future predictions.

When working with data problems that deal with very large sequences, traditional RNNs present a problem called the vanishing gradient problem. This is where the gradients become extremely small so that they no longer contribute to the learning process of the network, which typically occurs in the earlier layers of the network, causing the network to be unable to have a long-term memory.

In order to solve this problem, the LSTM network was developed. This network architecture is capable of storing two types of memory—hence its name. Additionally, the mathematical calculations that occur in this network allow it to forget information by only storing the relevant information from the past.

Finally, a very trendy NLP task was explained: sentiment analysis. In this task, it is important to understand the sentiment behind a text extraction. This is a very difficult problem for machines, considering that humans can use many different words and forms of expressions (for example, sarcasm) to describe the sentiment behind an event. However, thanks to the increase of social media usage, which has created a need to process text data faster, this problem has become very popular among large companies that have invested time and money to create several approximations in order to solve it, as shown in the final activity of this chapter.

Now that you have gone through all the chapters in this book, you have a wide understanding of different deep neural network architectures that can be used to solve a wide variety of data problems using PyTorch. The architectures that were explained in this book can also be adapted to solve other data problems.

.

APPENDIX

CHAPTER 1: INTRODUCTION TO DEEP LEARNING AND PYTORCH

ACTIVITY 1.01: CREATING A SINGLE-LAYER NEURAL NETWORK

SOLUTION

1. Import the required libraries, including pandas, for importing a CSV file:

```
import pandas as pd
import torch
import torch.nn as nn
import matplotlib.pyplot as plt
```

2. Read the CSV file containing the dataset:

```
data = pd.read_csv("SomervilleHappinessSurvey2015.csv")
```

3. Separate the input features from the target. Note that the target is located in the first column of the CSV file. Convert the values into tensors, making sure the values are converted into floats:

```
x = torch.tensor(data.iloc[:,1:].values).float()
y = torch.tensor(data.iloc[:,:1].values).float()
```

4. Define the architecture of the model and store it in a variable named **model**. Remember to create a single-layer model:

```
model = nn.Sequential(nn.Linear(6, 1),
                      nn.Sigmoid())
```

5. Define the loss function to be used. Use the MSE loss function:

```
loss_function = torch.nn.MSELoss()
```

6. Define the optimizer of your model. Use the Adam optimizer and a learning rate of **0.01**:

```
optimizer = torch.optim.Adam(model.parameters(), lr=0.01)
```

7. Run the optimization for 100 iterations. Every 10 iterations, print and save the loss value:

```
losses = []

for i in range(100):
    y_pred = model(x)
    loss = loss_function(y_pred, y)
```

```
losses.append(loss.item())
optimizer.zero_grad()
loss.backward()
optimizer.step()

if i%10 == 0:
    print(loss.item())
```

The final loss should be approximately **0.24**.

8. Make a line plot to display the loss value for each iteration step:

```
plt.plot(range(0,100), losses)
plt.show()
```

The resulting plot should look as follows:

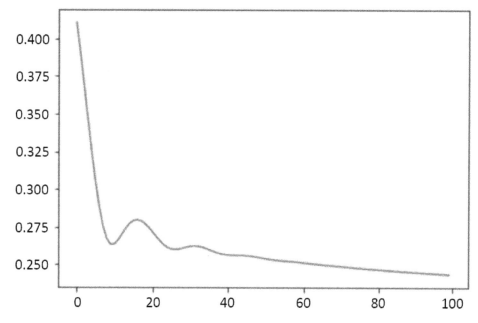

Figure 1.4: Loss function throughout the training process

This means that the training process is able to minimize the loss function, which means that the resulting model will likely be capable of mapping the relationship between the satisfaction of citizens with the city services and whether they are happy with the administration.

> **NOTE**
>
> To access the source code for this specific section, please refer to https://packt.live/2ZufWil.
>
> You can also run this example online at https://packt.live/2BZhyZF. You must execute the entire Notebook in order to get the desired result.

CHAPTER 2: BUILDING BLOCKS OF NEURAL NETWORKS

ACTIVITY 2.01: PERFORMING DATA PREPARATION

SOLUTION

1. Import the required libraries:

```
import pandas as pd
```

2. Using pandas, load the .csv file:

```
data = pd.read_csv("YearPredictionMSD.csv", nrows=50000)
data.head()
```

> **NOTE**
>
> To avoid memory limitations, use the **nrows** argument when reading the text file in order to read a smaller section of the entire dataset. In the preceding example, we are reading the first 50,000 rows.

The output is as follows:

	0	1	2	3	4	5	6	7	8	9	...	81
0	2001	49.94357	21.47114	73.07750	8.74861	-17.40628	-13.09905	-25.01202	-12.23257	7.83089	...	13.01620
1	2001	48.73215	18.42930	70.32679	12.94636	-10.32437	-24.83777	8.76630	-0.92019	18.76548	...	5.66812
2	2001	50.95714	31.85602	55.81851	13.41693	-6.57898	-18.54940	-3.27872	-2.35035	16.07017	...	3.03800
3	2001	48.24750	-1.89837	36.29772	2.58776	0.97170	-26.21683	5.05097	-10.34124	3.55005	...	34.57337
4	2001	50.97020	42.20998	67.09964	8.46791	-15.85279	-16.81409	-12.48207	-9.37636	12.63699	...	9.92661

Figure 2.33: YearPredictionMSD.csv

3. Verify whether any qualitative data is present in the dataset:

```
cols = data.columns

num_cols = data._get_numeric_data().columns

list(set(cols) - set(num_cols))
```

The output should be an empty list, meaning there are no qualitative features.

4. Check for missing values.

    If you add an additional **sum ()** function to the line of code that was previously used for this purpose, you will get the sum of missing values in the entire dataset, without discriminating by column:

    ```
    data.isnull().sum().sum()
    ```

    The output should be **0**, meaning that none of the features contain missing values.

5. Check for outliers:

    ```
    outliers = {}
    for i in range(data.shape[1]):
        min_t = data[data.columns[i]].mean() \
                - (3 * data[data.columns[i]].std())
        max_t = data[data.columns[i]].mean() \
                + (3 * data[data.columns[i]].std())

        count = 0
        for j in data[data.columns[i]]:
            if j < min_t or j > max_t:
                count += 1

        percentage = count/data.shape[0]
        outliers[data.columns[i]] = "%.3f" % percentage


    print(outliers)
    ```

    The output dictionary should display that none of the features contain outliers that represent over 5% of the data.

6. Separate the features from the target data:

    ```
    X = data.iloc[:, 1:]
    Y = data.iloc[:, 0]
    ```

7. Rescale the features data using the standardization methodology:

```
X = (X - X.mean())/X.std()
X.head()
```

The output is as follows:

	1	2	3	4	5	6	7	8	9	10	...	81
0	1.082657	0.382437	1.841985	0.459652	-0.480074	-0.282606	-1.590785	-1.300854	0.378336	-0.683719	...	-0.086005
1	0.880874	0.321953	1.763666	0.717085	-0.165507	-1.188896	0.777905	0.122576	1.420531	0.401198	...	-0.316635
2	1.251484	0.588929	1.350579	0.745944	0.000857	-0.703401	-0.066747	-0.057380	1.163637	-0.090081	...	-0.399185
3	0.800148	-0.082240	0.794774	0.081829	0.336246	-1.295366	0.517369	-1.062869	-0.029679	-1.282306	...	0.590596
4	1.253660	0.794806	1.671781	0.442438	-0.411071	-0.569426	-0.712128	-0.941459	0.836414	-0.160630	...	-0.182976

Figure 2.34: Rescaled features data

8. Split the data into three sets: training, validation, and test. Use the approach of your preference:

```
from sklearn.model_selection import train_test_split

X_shuffle = X.sample(frac=1, random_state=0)
Y_shuffle = Y.sample(frac=1, random_state=0)

x_new, x_test, \
y_new, y_test = train_test_split(X_shuffle, \
                                 Y_shuffle, \
                                 test_size=0.2, \
                                 random_state=0)
dev_per = x_test.shape[0]/x_new.shape[0]

x_train, x_dev, \
y_train, y_dev = train_test_split(x_new, \
                                  y_new, \
                                  test_size=dev_per, \
                                  random_state=0)
```

9. Print the resulting shapes as follows:

```
print(x_train.shape, y_train.shape)
print(x_dev.shape, y_dev.shape)
print(x_test.shape, y_test.shape)
```

The output should be as follows:

```
(30000, 90) (30000, )
(10000, 90) (10000, )
(10000, 90) (10000, )
```

> **NOTE**
>
> To access the source code for this specific section, please refer to
> https://packt.live/31ukVTj.
>
> You can also run this example online at https://packt.live/3dLWMdd.
> You must execute the entire Notebook in order to get the desired result.

ACTIVITY 2.02: DEVELOPING A DEEP LEARNING SOLUTION FOR A REGRESSION PROBLEM

SOLUTION

1. Import the required libraries:

```
import torch
import torch.nn as nn
```

2. Split the features from the targets for all three sets of data that we created in the previous activity. Convert the DataFrames into tensors:

```
x_train = torch.tensor(x_train.values).float()
y_train = torch.tensor(y_train.values).float()

x_dev = torch.tensor(x_dev.values).float()
y_dev = torch.tensor(y_dev.values).float()

x_test = torch.tensor(x_test.values).float()
y_test = torch.tensor(y_test.values).float()
```

3. Define the architecture of the network. Feel free to try different combinations for the number of layers and the number of units per layer:

```
model = nn.Sequential(nn.Linear(x_train.shape[1], 10), \
                      nn.ReLU(), \
                      nn.Linear(10, 7), \
                      nn.ReLU(), \
                      nn.Linear(7, 5), \
                      nn.ReLU(), \
                      nn.Linear(5, 1))
```

4. Define the loss function and the optimizer algorithm:

```
loss_function = torch.nn.MSELoss()
optimizer = torch.optim.Adam(model.parameters(), lr=0.01)
```

5. Use a **for** loop to train the network for 3,000 iteration steps:

```
for i in range(3000):
    y_pred = model(x_train).squeeze()
    loss = loss_function(y_pred, y_train)

    optimizer.zero_grad()
    loss.backward()
    optimizer.step()

    if i%250 == 0:
        print(i, loss.item())
```

6. Test your model by performing a prediction on the first instance of the test set and comparing it with the ground truth:

```
pred = model(x_test[0])
print("Ground truth:", y_test[0].item(), \
      "Prediction:", pred.item())
```

Your output should look similar to this:

```
Ground truth: 1995.0 Prediction: 1998.0279541015625
```

> **NOTE**
>
> To access the source code for this specific section, please refer to
> https://packt.live/2CUDSnP.
>
> You can also run this example online at https://packt.live/3eQ1yl2.
> You must execute the entire Notebook in order to get the desired result.

CHAPTER 3: A CLASSIFICATION PROBLEM USING DNNS

ACTIVITY 3.01: BUILDING AN ANN

Solution:

1. Import the following libraries:

```
import pandas as pd
import numpy as np
from sklearn.model_selection import train_test_split
from sklearn.utils import shuffle
from sklearn.metrics import accuracy_score
import torch
from torch import nn, optim
import torch.nn.functional as F
import matplotlib.pyplot as plt
torch.manual_seed(0)
```

2. Read the previously prepared dataset, which should have been named
dccc_prepared.csv:

```
data = pd.read_csv("dccc_prepared.csv")
data.head()
```

The output should be as follows:

	LIMIT_BAL	EDUCATION	MARRIAGE	AGE	PAY_0	PAY_2	PAY_3	PAY_4	PAY_5	PAY_6	...	BILL_AMT4
0	0.080808	0.333333	0.666667	0.224138	0.2	0.2	0.2	0.2	0.2	0.2	...	0.173637
1	0.040404	0.333333	0.333333	0.275862	0.2	0.2	0.2	0.2	0.2	0.2	...	0.186809
2	0.040404	0.333333	0.333333	0.620690	0.1	0.2	0.1	0.2	0.2	0.2	...	0.179863
3	0.040404	0.166667	0.666667	0.275862	0.2	0.2	0.2	0.2	0.2	0.2	...	0.178407
4	0.494949	0.166667	0.666667	0.137931	0.2	0.2	0.2	0.2	0.2	0.2	...	0.671310

Figure 3.14: dccc_prepared.csv

3. Separate the features from the target:

```
X = data.iloc[:,:-1]
y = data["default payment next month"]
```

4. Using scikit-learn's **train_test_split** function, split the dataset into training, validation, and testing sets. Use a 60:20:20 split ratio. Set **random_state** to 0:

```
X_new, X_test, \
y_new, y_test = train_test_split(X, y, test_size=0.2, \
                                 random_state=0)

dev_per = X_test.shape[0]/X_new.shape[0]
X_train, X_dev, \
y_train, y_dev = train_test_split(X_new, y_new, \
                                  test_size=dev_per, \
                                  random_state=0)
```

You can print the final shapes of each of the set using the following code:

```
print("Training sets:",X_train.shape, y_train.shape)
print("Validation sets:",X_dev.shape, y_dev.shape)
print("Testing sets:",X_test.shape, y_test.shape)
```

The final shapes of each of the sets are shown here:

```
Training sets: (28036, 22) (28036,)
Validation sets: (9346, 22) (9346,)
Testing sets: (9346, 22) (9346,)
```

5. Convert the validation and testing sets into tensors, bearing in mind that the features' matrices should be of the float type, while the target matrices should not. Leave the training sets unconverted for the moment as they will undergo further transformation:

```
X_dev_torch = torch.tensor(X_dev.values).float()
y_dev_torch = torch.tensor(y_dev.values)
X_test_torch = torch.tensor(X_test.values).float()
y_test_torch = torch.tensor(y_test.values)
```

6. Build a custom module class for defining the layers of the network. Include a forward function that specifies the activation functions that will be applied to the output of each layer. Use **ReLU** for all the layers, except for the output, where you should use **log_softmax**:

```
class Classifier(nn.Module):
    def __init__(self, input_size):
        super().__init__()
        self.hidden_1 = nn.Linear(input_size, 10)
        self.hidden_2 = nn.Linear(10, 10)
        self.hidden_3 = nn.Linear(10, 10)
        self.output = nn.Linear(10, 2)

    def forward(self, x):
        z = F.relu(self.hidden_1(x))
        z = F.relu(self.hidden_2(z))
        z = F.relu(self.hidden_3(z))
        out = F.log_softmax(self.output(z), dim=1)

        return out
```

7. Instantiate the model and define all the variables required to train the model. Set the number of epochs to **50** and the batch size to **128**. Use a learning rate of **0.001**:

```
model = Classifier(X_train.shape[1])
criterion = nn.NLLLoss()
optimizer = optim.Adam(model.parameters(), lr=0.001)

epochs = 50
batch_size = 128
```

8. Train the network using the training sets' data. Use the validation sets to measure performance. To do this, save the loss and the accuracy for both the training and validation sets in each epoch:

```
train_losses, dev_losses, \
train_acc, dev_acc = [], [], [], []

for e in range(epochs):
    X_, y_ = shuffle(X_train, y_train)
    running_loss = 0
```

```python
running_acc = 0
iterations = 0

for i in range(0, len(X_), batch_size):
    iterations += 1
    b = i + batch_size
    X_batch = torch.tensor(X_.iloc[i:b,:].values).float()
    y_batch = torch.tensor(y_.iloc[i:b].values)

    pred = model(X_batch)
    loss = criterion(pred, y_batch)
    optimizer.zero_grad()
    loss.backward()
    optimizer.step()

    running_loss += loss.item()
    ps = torch.exp(pred)
    top_p, top_class = ps.topk(1, dim=1)
    running_acc += accuracy_score(y_batch, top_class)

dev_loss = 0
acc = 0
with torch.no_grad():
    pred_dev = model(X_dev_torch)
    dev_loss = criterion(pred_dev, y_dev_torch)

    ps_dev = torch.exp(pred_dev)
    top_p, top_class_dev = ps_dev.topk(1, dim=1)
    acc = accuracy_score(y_dev_torch, top_class_dev)

train_losses.append(running_loss/iterations)
dev_losses.append(dev_loss)
train_acc.append(running_acc/iterations)
dev_acc.append(acc)

print("Epoch: {}/{}.. ".format(e+1, epochs),\
        "Training Loss: {:.3f}.. "\
```

```
                    .format(running_loss/iterations),\
                    "Validation Loss: {:.3f}.. ".format(dev_loss), \
                    "Training Accuracy: {:.3f}.. "\
                    .format(running_acc/iterations), \
                    "Validation Accuracy: {:.3f}".format(acc))
```

9. Plot the loss of both sets:

```
fig = plt.figure(figsize=(15, 5))
plt.plot(train_losses, label='Training loss')
plt.plot(dev_losses, label='Validation loss')
plt.legend(frameon=False, fontsize=15)
plt.show()
```

The resulting plot should look similar to the one shown here, albeit with some differences, considering that shuffling the training data may derive slightly different results:

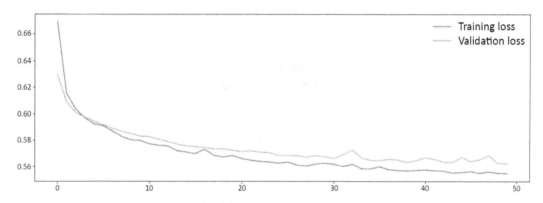

Figure 3.15: A plot displaying the training and validation losses

10. Plot the accuracy of both sets:

```
fig = plt.figure(figsize=(15, 5))
plt.plot(train_acc, label="Training accuracy")
plt.plot(dev_acc, label="Validation accuracy")
plt.legend(frameon=False, fontsize=15)
plt.show()
```

Here is the plot that's derived from this code snippet:

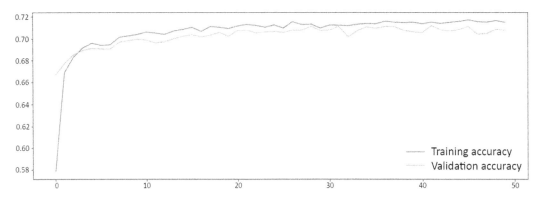

Figure 3.16: A plot displaying the accuracy of the sets

> **NOTE**
>
> To access the source code for this specific section, please refer to https://packt.live/2Vz6BoK.
>
> You can also run this example online at https://packt.live/2NNBuRS. You must execute the entire Notebook in order to get the desired result.

ACTIVITY 3.02: IMPROVING A MODEL'S PERFORMANCE

Solution:

1. Import the same libraries that you used in the previous activity:

```
import pandas as pd
import numpy as np
from sklearn.model_selection import train_test_split
from sklearn.utils import shuffle
from sklearn.metrics import accuracy_score
import torch
from torch import nn, optim
import torch.nn.functional as F
import matplotlib.pyplot as plt
torch.manual_seed(0)
```

2. Load the data and split the features from the target. Next, split the data into the three subsets (training, validation, and testing) using a 60:20:20 split ratio. Finally, convert the validation and testing sets into PyTorch tensors, just as you did in the previous activity:

```
data = pd.read_csv("dccc_prepared.csv")

X = data.iloc[:,:-1]
y = data["default payment next month"]

X_new, X_test, \
y_new, y_test = train_test_split(X, y, test_size=0.2, \
                                 random_state=0)

dev_per = X_test.shape[0]/X_new.shape[0]

X_train, X_dev, \
y_train, y_dev = train_test_split(X_new, y_new, \
                                  test_size=dev_per, \
                                  random_state=0)

X_dev_torch = torch.tensor(X_dev.values).float()
y_dev_torch = torch.tensor(y_dev.values)
X_test_torch = torch.tensor(X_test.values).float()
y_test_torch = torch.tensor(y_test.values)
```

3. Considering that the model is suffering from high bias, the focus should be on increasing the number of epochs or increasing the size of the network by adding additional layers or units to each layer. The aim should be to approximate the accuracy over the validation set to 80%.

Following this, the best-performing model is displayed, which is achieved after several fine-tuning attempts. First, the model architecture and forward pass are defined, as can be seen in the following code snippet:

```
class Classifier(nn.Module):
    def __init__(self, input_size):
        super().__init__()
        self.hidden_1 = nn.Linear(input_size, 100)
        self.hidden_2 = nn.Linear(100, 100)
```

```
        self.hidden_3 = nn.Linear(100, 50)
        self.hidden_4 = nn.Linear(50,50)
        self.output = nn.Linear(50, 2)

        self.dropout = nn.Dropout(p=0.1)

    def forward(self, x):
        z = self.dropout(F.relu(self.hidden_1(x)))
        z = self.dropout(F.relu(self.hidden_2(z)))
        z = self.dropout(F.relu(self.hidden_3(z)))
        z = self.dropout(F.relu(self.hidden_4(z)))
        out = F.log_softmax(self.output(z), dim=1)

        return out
```

Next, the different parameters of the training process are defined. This includes the loss function, the optimization algorithm, the batch size, and the number of epochs, as seen in the following code:

```
model = Classifier(X_train.shape[1])
criterion = nn.NLLLoss()
optimizer = optim.Adam(model.parameters(), lr=0.001)

epochs = 4000
batch_size = 128
```

Finally, the training process is handled, as per the following code snippet:

```
train_losses, dev_losses, train_acc, dev_acc= [], [], [], []
x_axis = []

for e in range(1, epochs + 1):
    X_, y_ = shuffle(X_train, y_train)
    running_loss = 0
    running_acc = 0
    iterations = 0

    for i in range(0, len(X_), batch_size):
        iterations += 1
        b = i + batch_size
```

```python
            X_batch = torch.tensor(X_.iloc[i:b,:].values).float()
            y_batch = torch.tensor(y_.iloc[i:b].values)

            log_ps = model(X_batch)
            loss = criterion(log_ps, y_batch)
            optimizer.zero_grad()
            loss.backward()
            optimizer.step()

            running_loss += loss.item()
            ps = torch.exp(log_ps)
            top_p, top_class = ps.topk(1, dim=1)
            running_acc += accuracy_score(y_batch, top_class)

    dev_loss = 0
    acc = 0

    with torch.no_grad():
        model.eval()
        log_dev = model(X_dev_torch)
        dev_loss = criterion(log_dev, y_dev_torch)

        ps_dev = torch.exp(log_dev)
        top_p, top_class_dev = ps_dev.topk(1, dim=1)
        acc = accuracy_score(y_dev_torch, top_class_dev)

    model.train()

    if e%50 == 0 or e == 1:
        x_axis.append(e)

        train_losses.append(running_loss/iterations)
        dev_losses.append(dev_loss)
        train_acc.append(running_acc/iterations)
        dev_acc.append(acc)

        print("Epoch: {}/{}.. ".format(e, epochs), \
              "Training Loss: {:.3f}.. "\
```

```
.format(running_loss/iterations), \
"Validation Loss: {:.3f}.. ".format(dev_loss),\
"Training Accuracy: {:.3f}.. "\
.format(running_acc/iterations), \
"Validation Accuracy: {:.3f}".format(acc))
```

> **NOTE**
>
> The accompanying Jupyter Notebook for this activity can be found in the GitHub repository that was shared previously. There, you will find the different attempts at fine-tuning the model, along with their results. The best-performing model can be found at the end of the notebook.

4. Plot the loss and accuracy for both sets of data:

> **NOTE**
>
> Keep in mind that the results presented here will not match your results exactly. This is mainly due to the shuffling function that was used while training the networks.

Use the following code to plot loss:

```
fig = plt.figure(figsize=(15, 5))
plt.plot(x_axis,train_losses, label='Training loss')
plt.plot(x_axis, dev_losses, label='Validation loss')
plt.legend(frameon=False , fontsize=15)
plt.show()
```

Running the preceding code displays the following plot:

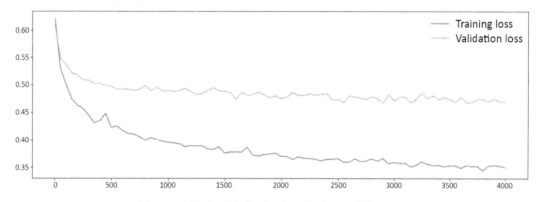

Figure 3.17: A plot displaying the loss of the sets

Use the following code to plot accuracy:

```
fig = plt.figure(figsize=(15, 5))
plt.plot(x_axis, train_acc, label="Training accuracy")
plt.plot(x_axis, dev_acc, label="Validation accuracy")
plt.legend(frameon=False , fontsize=15)
plt.show()
```

Running the preceding code displays the following plot:

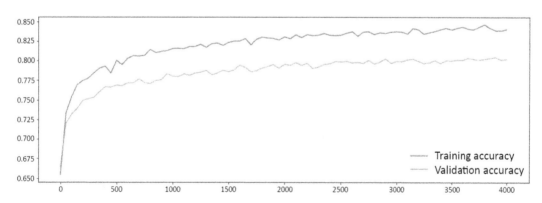

Figure 3.18: A plot displaying the accuracy of the sets

5. Using the best-performing model, perform a prediction over the testing set (which should not have been used during the fine-tuning process). Compare the prediction with the ground truth by calculating the accuracy of the model over this set:

```
model.eval()
test_pred = model(X_test_torch)
test_pred = torch.exp(test_pred)
top_p, top_class_test = test_pred.topk(1, dim=1)
acc_test = accuracy_score(y_test_torch, top_class_test)
print(acc_test)
```

The accuracy that was obtained through the model architecture and the parameters that were defined here should be around 80%.

> **NOTE**
>
> To access the source code for this specific section, please refer to https://packt.live/2Bs42hh.
>
> This section does not currently have an online interactive example, and will need to be run locally.

ACTIVITY 3.03: MAKING USE OF YOUR MODEL

SOLUTION

1. Open the Jupyter Notebook that you used for the previous activity.

2. Copy the class containing the architecture of your best-performing model and save it in a Python file. Make sure that you import PyTorch's required libraries and modules. Name it **final_model.py**.

The file should look as follows:

```python
import torch
from torch import nn, optim
import torch.nn.functional as F

class Classifier(nn.Module):
    def __init__(self, input_size):
        super().__init__()
        self.hidden_1 = nn.Linear(input_size, 100)
        self.hidden_2 = nn.Linear(100, 100)
        self.hidden_3 = nn.Linear(100, 50)
        self.hidden_4 = nn.Linear(50,50)
        self.output = nn.Linear(50, 2)

        self.dropout = nn.Dropout(p=0.1)
        #self.dropout_2 = nn.Dropout(p=0.1)

    def forward(self, x):
        z = self.dropout(F.relu(self.hidden_1(x)))
        z = self.dropout(F.relu(self.hidden_2(z)))
        z = self.dropout(F.relu(self.hidden_3(z)))
        z = self.dropout(F.relu(self.hidden_4(z)))
        out = F.log_softmax(self.output(z), dim=1)

        return out
```

Figure 3.19: A screenshot of final_model.py

3. In the Jupyter Notebook, save the best-performing model. Make sure to save the information pertaining to the input units, along with the parameters of the model. Name it **checkpoint.pth**:

```
checkpoint = {"input": X_train.shape[1], \
              "state_dict": model.state_dict()}
torch.save(checkpoint, "checkpoint.pth")
```

4. Open a new Jupyter Notebook.

5. Import PyTorch, as well as the Python file we created in *Step 2*:

```
import torch
import final_model
```

6. Create a function that loads the model:

```
def load_model_checkpoint(path):
    checkpoint = torch.load(path)

    model = final_model.Classifier(checkpoint["input"])

    model.load_state_dict(checkpoint["state_dict"])

    return model

model = load_model_checkpoint("checkpoint.pth")
```

7. Perform a prediction by inputting the following tensor into your model:

```
example = torch.tensor([[0.0606, 0.5000, 0.3333, 0.4828, \
                         0.4000, 0.4000, 0.4000, 0.4000, \
                         0.4000, 0.4000, 0.1651, 0.0869, \
                         0.0980, 0.1825, 0.1054, 0.2807, \
                         0.0016, 0.0000, 0.0033, 0.0027, \
                         0.0031, 0.0021]]).float()
```

```
pred = model(example)
pred = torch.exp(pred)
top_p, top_class_test = pred.topk(1, dim=1)
```

By printing **top_class_test**, we obtain the prediction of the model, which, in this case, is equal to **1** (yes).

8. Convert the model using the JIT module:

```
traced_script = torch.jit.trace(model, example, \
                        check_trace=False)
```

9. Perform a prediction by inputting the same tensor as in *Step 7* to the traced script of your model:

```
prediction = traced_script(example)
prediction = torch.exp(prediction)
top_p_2, top_class_test_2 = prediction.topk(1, dim=1)
```

By printing **top_class_test_2**, we get the prediction from the traced script representation of your model, which again is equal to **1** (yes).

10. Open a new Jupyter Notebook and import the required libraries to create an API using Flask, as well as the libraries to load the saved model:

```
import flask
from flask import request
import torch
import final_model
```

11. Initialize the Flask app:

```
app = flask.Flask(__name__)
app.config["DEBUG"] = True
```

12. Define a function that loads the saved model and then instantiate the model:

```
def load_model_checkpoint(path):
    checkpoint = torch.load(path)

    model = final_model.Classifier(checkpoint["input"])

    model.load_state_dict(checkpoint["state_dict"])

    return model

model = load_model_checkpoint("checkpoint.pth")
```

13. Define the route of the API to **/prediction** and set the method to **POST**. Then, define the function that will receive the **POST** data and feed it to the model to perform a prediction:

```
@app.route('/prediction', methods=['POST'])
def prediction():

    body = request.get_json()

    example = torch.tensor(body['data']).float()

    pred = model(example)
    pred = torch.exp(pred)
    _, top_class_test = pred.topk(1, dim=1)
    top_class_test = top_class_test.numpy()

    return {"status":"ok", "result":int(top_class_test[0][0])}
```

14. Run the Flask app:

```
app.run(debug=True, use_reloader=False)
```

Using Postman, a platform that was created for API development, it is possible to test the API. To submit a successful request to Postman, the header should have **Content-Type** equal to **application/json**. The resulting output should be as follows:

Figure 3.20: A screenshot of the app after running it

> **NOTE**
>
> To access the source code for this specific section, please refer to
> https://packt.live/2NHkddn.
>
> This section does not currently have an online interactive example, and will
> need to be run locally.

CHAPTER 4: CONVOLUTIONAL NEURAL NETWORKS

ACTIVITY 4.01: BUILDING A CNN FOR AN IMAGE CLASSIFICATION PROBLEM

SOLUTION

1. Import the required libraries:

```
import numpy as np
import torch
from torch import nn, optim
import torch.nn.functional as F
from torchvision import datasets
import torchvision.transforms as transforms
from torch.utils.data.sampler import SubsetRandomSampler
from sklearn.metrics import accuracy_score
import matplotlib.pyplot as plt
```

2. Set the transformations to be performed on the data, which will be converting the data into tensors and normalizing the pixel values:

```
transform = \
    transforms.Compose([transforms.ToTensor(), \
                        transforms.Normalize((0.5, 0.5, 0.5),\
                                             (0.5, 0.5, 0.5))])
```

3. Set a batch size of 100 images and download both the training and testing data from the **CIFAR10** dataset:

```
batch_size = 100

train_data = datasets.CIFAR10('data', train=True, \
                              download=True, \
                              transform=transform)

test_data = datasets.CIFAR10('data', train=False, \
                             download=True, \
                             transform=transform)
```

The preceding code downloads both the training and testing datasets that are available through PyTorch's **torchvision** package. The datasets are transformed as per the transformations defined in the previous step.

4. Using a validation size of 20%, define the training and validation sampler that will be used to divide the dataset into those two sets:

```
dev_size = 0.2
idx = list(range(len(train_data)))
np.random.shuffle(idx)
split_size = int(np.floor(dev_size * len(train_data)))
train_idx, dev_idx = idx[split_size:], idx[:split_size]

train_sampler = SubsetRandomSampler(train_idx)
dev_sampler = SubsetRandomSampler(dev_idx)
```

In order to split the training set into two sets (training and validation), a list of indexes is defined for each of the sets, which can then be randomly sampled using the **SubsetRandomSampler** function.

5. Use the **DataLoader()** function to define the batches of each set of data to be used:

```
train_loader = \
torch.utils.data.DataLoader(train_data, \
                            batch_size=batch_size, \
                            sampler=train_sampler)
dev_loader = \
torch.utils.data.DataLoader(train_data, \
                            batch_size=batch_size, \
                            sampler=dev_sampler)

test_loader = \
torch.utils.data.DataLoader(test_data, \
                            batch_size=batch_size)
```

PyTorch's **DataLoader** function is used to allow the creation of batches that will be fed to the model during the training, validation, and testing phases of the development process.

6. Define the architecture of your network. Use the following information to do so:

Conv1: A convolutional layer that takes the colored image as input and passes it through 10 filters of size 3. Both the padding and the stride should be set to 1.

Conv2: A convolutional layer that passes the input data through 20 filters of size 3. Both the padding and the stride should be set to 1.

Conv3: A convolutional layer that passes the input data through 40 filters of size 3. Both the padding and the stride should be set to 1.

Use the ReLU activation function after each convolutional layer.

Use a pooling layer after each convolutional layer, with a filter size and stride of 2.

Use a dropout term set to 20%, after flattening the image.

Linear1: A fully connected layer that receives the flattened matrix from the previous layer as input and generates an output of 100 units. Use the ReLU activation function for this layer. The dropout term here is set to 20%.

Linear2: A fully connected layer that generates 10 outputs, one for each class label. Use the **log_softmax** activation function for the output layer:

```python
class CNN(nn.Module):
    def __init__(self):
        super(CNN, self).__init__()
        self.conv1 = nn.Conv2d(3, 10, 3, 1, 1)
        self.conv2 = nn.Conv2d(10, 20, 3, 1, 1)
        self.conv3 = nn.Conv2d(20, 40, 3, 1, 1)
        self.pool = nn.MaxPool2d(2, 2)

        self.linear1 = nn.Linear(40 * 4 * 4, 100)
        self.linear2 = nn.Linear(100, 10)
        self.dropout = nn.Dropout(0.2)

    def forward(self, x):
        x = self.pool(F.relu(self.conv1(x)))
        x = self.pool(F.relu(self.conv2(x)))
        x = self.pool(F.relu(self.conv3(x)))

        x = x.view(-1, 40 * 4 * 4)
        x = self.dropout(x)
        x = F.relu(self.linear1(x))
        x = self.dropout(x)
        x = F.log_softmax(self.linear2(x), dim=1)

        return x
```

The preceding code snippet consists of a class where the network architecture is defined (the **__init__** method), as well as the steps that are followed during the forward pass of the information (the **forward** method).

7. Define all of the parameters that are required to train your model. Set the number of epochs to **50**:

```
model = CNN()
loss_function = nn.NLLLoss()
optimizer = optim.Adam(model.parameters(), lr=0.001)
epochs = 50
```

The optimizer that we selected for this exercise is Adam. Also, the negative log-likelihood is used as the loss function, as in the previous chapter of this book.

If your machine has a GPU available, the instantiation of the model should be done as follows:

```
model = CNN().to("cuda")
```

8. Train your network and be sure to save the values for the loss and accuracy of both the training and validation sets:

```
train_losses, dev_losses, train_acc, dev_acc= [], [], [], []
x_axis = []

# For loop through the epochs
for e in range(1, epochs+1):
    losses = 0
    acc = 0
    iterations = 0

    model.train()

    """
    For loop through the batches (created using
    the train loader)
    """
    for data, target in train_loader:
        iterations += 1

        # Forward and backward pass of the training data
        pred = model(data)
```

```python
        loss = loss_function(pred, target)
        optimizer.zero_grad()
        loss.backward()
        optimizer.step()

        losses += loss.item()
        p = torch.exp(pred)
        top_p, top_class = p.topk(1, dim=1)
        acc += accuracy_score(target, top_class)

dev_losss = 0
dev_accs = 0
iter_2 = 0

# Validation of model for given epoch
if e%5 == 0 or e == 1:
    x_axis.append(e)

    with torch.no_grad():
        model.eval()

        """
        For loop through the batches of
        the validation set
        """
        for data_dev, target_dev in dev_loader:
            iter_2 += 1

            dev_pred = model(data_dev)
            dev_loss = loss_function(dev_pred, target_dev)
            dev_losss += dev_loss.item()

            dev_p = torch.exp(dev_pred)
            top_p, dev_top_class = dev_p.topk(1, dim=1)
            dev_accs += accuracy_score(target_dev, \
                                        dev_top_class)

    # Losses and accuracy are appended to be printed
    train_losses.append(losses/iterations)
    dev_losses.append(dev_losss/iter_2)
```

```
            train_acc.append(acc/iterations)
            dev_acc.append(dev_accs/iter_2)

        print("Epoch: {}/{}.. ".format(e, epochs), \
              "Training Loss: {:.3f}.. "\
              .format(losses/iterations), \
              "Validation Loss: {:.3f}.. "\
              .format(dev_losss/iter_2), \
              "Training Accuracy: {:.3f}.. "\
              .format(acc/iterations), \
              "Validation Accuracy: {:.3f}"\
              .format(dev_accs/iter_2))
```

If your machine has a GPU available, some modifications to the preceding code apply, as follows:

```
train_losses, dev_losses, train_acc, dev_acc= [], [], [], []
x_axis = []

# For loop through the epochs
for e in range(1, epochs+1):
    losses = 0
    acc = 0
    iterations = 0

    model.train()

    """
    For loop through the batches
    (created using the train loader)
    """
    for data, target in train_loader:
        iterations += 1

        # Forward and backward pass of the training data
        pred = model(data.to("cuda"))
        loss = loss_function(pred, target.to("cuda"))
        optimizer.zero_grad()
        loss.backward()
        optimizer.step()
```

```
        losses += loss.item()
        p = torch.exp(pred)
        top_p, top_class = p.topk(1, dim=1)
        acc += accuracy_score(target.to("cpu"), \
                top_class.to("cpu"))

    dev_losss = 0
    dev_accs = 0
    iter_2 = 0

    # Validation of model for given epoch
    if e%5 == 0 or e == 1:
        x_axis.append(e)

        with torch.no_grad():
            model.eval()

            """
            For loop through the batches of
            the validation set
            """
            for data_dev, target_dev in dev_loader:
                iter_2 += 1

                dev_pred = model(data_dev.to("cuda"))
                dev_loss = loss_function(dev_pred, \
                        target_dev.to("cuda"))
                dev_losss += dev_loss.item()

                dev_p = torch.exp(dev_pred)
                top_p, dev_top_class = dev_p.topk(1, dim=1)
                dev_accs += \
                accuracy_score(target_dev.to("cpu"), \
                        dev_top_class.to("cpu"))

        # Losses and accuracy are appended to be printed
        train_losses.append(losses/iterations)
        dev_losses.append(dev_losss/iter_2)
        train_acc.append(acc/iterations)
```

```
dev_acc.append(dev_accs/iter_2)

print("Epoch: {}/{}.. ".format(e, epochs), \
        "Training Loss: {:.3f}.. "\
        .format(losses/iterations), \
        "Validation Loss: {:.3f}.. "\
        .format(dev_losss/iter_2), \
        "Training Accuracy: {:.3f}.. "\
        .format(acc/iterations), \
        "Validation Accuracy: {:.3f}"\
        .format(dev_accs/iter_2))
```

9. Plot the loss and accuracy of both sets. To plot the loss, use the following code:

```
plt.plot(x_axis,train_losses, label='Training loss')
plt.plot(x_axis, dev_losses, label='Validation loss')
plt.legend(frameon=False)
plt.show()
```

The resulting plot should look similar to the following:

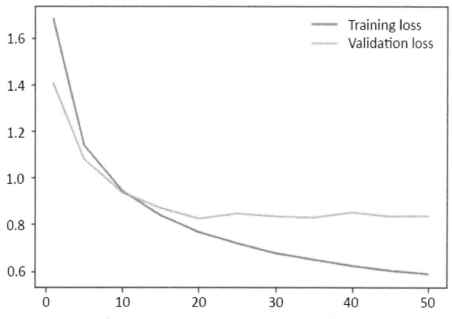

Figure 4.23: Resulting plot showing the loss of the sets

To plot the accuracy, use the following code:

```
plt.plot(x_axis, train_acc, label="Training accuracy")
plt.plot(x_axis, dev_acc, label="Validation accuracy")
plt.legend(frameon=False)
plt.show()
```

The plot should look similar to the following:

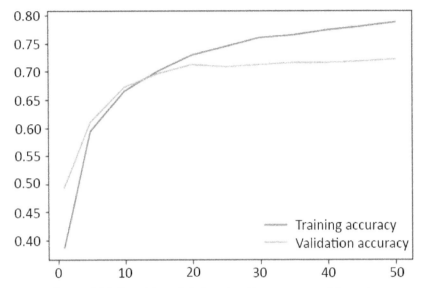

Figure 4.24: Resulting plot showing the accuracy of the sets

As can be seen, after the 15th epoch, overfitting starts to affect the model.

10. Check the model's accuracy on the testing set:

```
model.eval()
iter_3 = 0
acc_test = 0

for data_test, target_test in test_loader:
    iter_3 += 1
    test_pred = model(data_test)
    test_pred = torch.exp(test_pred)
    top_p, top_class_test = test_pred.topk(1, dim=1)
    acc_test += accuracy_score(target_test, top_class_test)

print(acc_test/iter_3)
```

Using the data loader we created previously, it is possible to perform the classification of images on the testing set data in order to estimate the model's accuracy on unseen data.

If your machine has a GPU available, some modifications to the preceding code apply, as follows:

```
model.eval()
iter_3 = 0
acc_test = 0

for data_test, target_test in test_loader:
    iter_3 += 1
    test_pred = model(data_test.to("cuda"))
    test_pred = torch.exp(test_pred)
    top_p, top_class_test = test_pred.topk(1, dim=1)
    acc_test += accuracy_score(target_test .to("cpu"), \
                        top_class_test .to("cpu"))


print(acc_test/iter_3)
```

The accuracy of the testing set is very similar to the accuracy that was achieved by the other two sets, which means that the model has the capability to perform equally well on unseen data; it should be around 72%.

> **NOTE**
>
> To access the source code for this specific section, please refer to https://packt.live/3gjvWuV.
>
> This section does not currently have an online interactive example, and will need to be run locally.
>
> To access the GPU version of this source code, please refer to https://packt.live/2BUGjGF. This version of the source code is not available as an online interactive example, and will need to be run locally with the GPU setup.

ACTIVITY 4.02: IMPLEMENTING DATA AUGMENTATION

SOLUTION

1. Duplicate the notebook from the previous activity.

    To complete this activity, no code will be altered besides the modification of the **transforms** value, as per the following step.

2. Change the definition of the **transform** variable so that it includes, in addition to normalizing and converting the data into tensors, the following transformations:

    For the training/validation sets, use a **RandomHorizontalFlip** function with a probability of 50% (**0.5**) and a **RandomGrayscale** function with a probability of 10% (**0.1**).

    For the testing set, do not add any other transformations:

```
transform = \
{"train": transforms.Compose([\
          transforms.RandomHorizontalFlip(0.5), \
          transforms.RandomGrayscale(0.1),\
          transforms.ToTensor(),\
          transforms.Normalize((0.5, 0.5, 0.5), \
                                (0.5, 0.5, 0.5))]),\

"test": transforms.Compose([\
        transforms.ToTensor(),\
        transforms.Normalize((0.5, 0.5, 0.5), \
                              (0.5, 0.5, 0.5))])}
```

3. Train the model for 100 epochs.

    If your machine has a GPU available, make sure to use the GPU version of the code to train the model.

The resulting plots for loss and accuracy on the training and validation sets should be similar to the ones shown here:

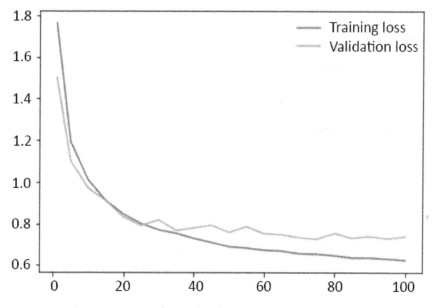

Figure 4.25: Resulting plot showing the loss of the sets

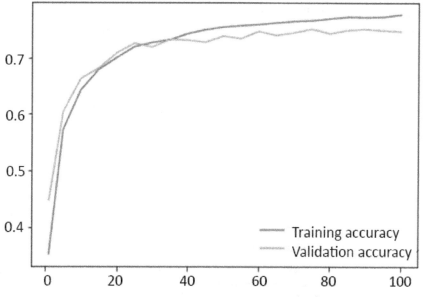

Figure 4.26: Resulting plot showing the accuracy of the sets

By adding data augmentation, it is possible to improve the performance of the model, as well as reduce the overfitting that was occurring.

4. Calculate the accuracy of the resulting model on the testing set.

The performance of the model on the testing set has gone up to around 75%.

> **NOTE**
>
> To access the source code for this specific section, please refer to https://packt.live/3ePcAND.
>
> This section does not currently have an online interactive example, and will need to be run locally.
>
> To access the GPU version of this source code, please refer to https://packt.live/38jpq4g. This version of the source code is not available as an online interactive example, and will need to be run locally with the GPU setup.

ACTIVITY 4.03: IMPLEMENTING BATCH NORMALIZATION

SOLUTION

1. Duplicate the notebook from the previous activity.

To complete this activity, no code will be altered besides the addition of some layers to the network architecture, as per the following step.

2. Add batch normalization to each convolutional layer, as well as to the first fully connected layer.

The resulting architecture of the network should be as follows:

```
class CNN(nn.Module):

    def __init__(self):
        super(Net, self).__init__()

        self.conv1 = nn.Conv2d(3, 10, 3, 1, 1)
        self.norm1 = nn.BatchNorm2d(10)
        self.conv2 = nn.Conv2d(10, 20, 3, 1, 1)
        self.norm2 = nn.BatchNorm2d(20)
        self.conv3 = nn.Conv2d(20, 40, 3, 1, 1)
        self.norm3 = nn.BatchNorm2d(40)
        self.pool = nn.MaxPool2d(2, 2)
```

```
        self.linear1 = nn.Linear(40 * 4 * 4, 100)
        self.norm4 = nn.BatchNorm1d(100)
        self.linear2 = nn.Linear(100, 10)
        self.dropout = nn.Dropout(0.2)

    def forward(self, x):
        x = self.pool(self.norm1(F.relu(self.conv1(x))))
        x = self.pool(self.norm2(F.relu(self.conv2(x))))
        x = self.pool(self.norm3(F.relu(self.conv3(x))))

        x = x.view(-1, 40 * 4 * 4)
        x = self.dropout(x)
        x = self.norm4(F.relu(self.linear1(x)))
        x = self.dropout(x)
        x = F.log_softmax(self.linear2(x), dim=1)

        return x
```

3. Train the model for 100 epochs.

   If your machine has a GPU available, make sure to use the GPU version of the code to train the model. The resulting plots of the loss and accuracy of the training and validation sets should be similar to the ones shown here:

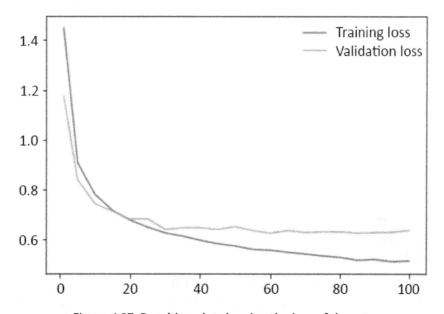

Figure 4.27: Resulting plot showing the loss of the sets

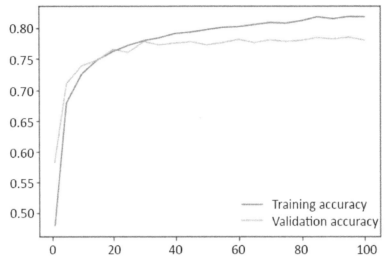

Figure 4.28: Resulting plot showing the loss of the sets

Although overfitting was introduced to the model again, we can see that the performance of both sets has gone up.

> **NOTE**
>
> Although it is not explored in this chapter, an ideal step would be to add dropout to the architecture of the network in order to reduce high variance. Feel free to try it to see if you are able to improve the performance even more.

4. Calculate the accuracy of the resulting model on the testing set.

    The performance of the model on the testing set has gone up to around 78%.

> **NOTE**
>
> To access the source code for this specific section, please refer to https://packt.live/31sSR2G.
>
> This section does not currently have an online interactive example, and will need to be run locally.
>
> To access the GPU version of this source code, please refer to https://packt.live/3eVgp4g. This version of the source code is not available as an online interactive example, and will need to be run locally with the GPU setup.

CHAPTER 5: STYLE TRANSFER

ACTIVITY 5.01: PERFORMING STYLE TRANSFER

SOLUTION

1. Import the required libraries:

```
import numpy as np
import torch
from torch import nn, optim
from PIL import Image
import matplotlib.pyplot as plt
from torchvision import transforms, models
```

If your machine has a GPU available, make sure to define a variable named **device** that will help to allocate some variables to the GPU, as follows:

```
device = "cuda"
```

2. Specify the transformations to be performed over the input images. Be sure to resize them to the same size, convert them into tensors, and normalize them:

```
imsize = 224

loader = \
transforms.Compose([transforms.Resize(imsize), \
                    transforms.ToTensor(),\
                    transforms.Normalize((0.485, 0.456, 0.406), \
                                         (0.229, 0.224, 0.225))])
```

3. Define an image loader function. It should open the image and load it. Call the image loader function to load both input images:

```
def image_loader(image_name):
    image = Image.open(image_name)
    image = loader(image).unsqueeze(0)
    return image

content_img = image_loader("images/landscape.jpg")
style_img = image_loader("images/monet.jpg")
```

If your machine has a GPU available, use the following code snippet instead:

```
def image_loader(image_name):
    image = Image.open(image_name)
    image = loader(image).unsqueeze(0)
    return image


content_img = image_loader("images/landscape.jpg").to(device)
style_img = image_loader("images/monet.jpg").to(device)
```

4. To be able to display the images, set the transformations to revert the normalization of the images and to convert the tensors into **PIL** images:

```
unloader = transforms.Compose([\
            transforms.Normalize((-0.485/0.229, \
                                    -0.456/0.224, \
                                    -0.406/0.225), \
                                    (1/0.229, 1/0.224, 1/0.225)),\
            transforms.ToPILImage()])
```

5. Create a function (**tensor2image**) that's capable of performing the previous transformation over tensors. Call the function for both images and plot the results:

```
def tensor2image(tensor):
    image = tensor.clone()
    image = image.squeeze(0)
    image = unloader(image)
    return image


plt.figure()
plt.imshow(tensor2image(content_img))
plt.title("Content Image")
plt.show()

plt.figure()
plt.imshow(tensor2image(style_img))
plt.title("Style Image")
plt.show()
```

If your machine has a GPU available, use the following code snippet instead:

```
def tensor2image(tensor):
    image = tensor.to("cpu").clone()
    image = image.squeeze(0)
    image = unloader(image)
    return image


plt.figure()
plt.imshow(tensor2image(content_img))
plt.title("Content Image")
plt.show()

plt.figure()
plt.imshow(tensor2image(style_img))
plt.title("Style Image")
plt.show()
```

6. Load the VGG-19 model:

```
model = models.vgg19(pretrained=True).features
for param in model.parameters():
    param.requires_grad_(False)
```

If your machine has a GPU available, make sure to allocate the variable containing your model to the GPU, as follows:

```
model.to(device)
```

7. Create a dictionary for mapping the index of the relevant layers (keys) to a name (values). Then, create a function to extract the feature maps of the relevant layers. Use them to extract the features of both input images.

The following function should extract the features of a given image for each of the relevant layers:

```
relevant_layers = {'0': 'conv1_1', '5': 'conv2_1', \
                   '10': 'conv3_1', '19': 'conv4_1', \
                   '21': 'conv4_2', '28': 'conv5_1'}

def features_extractor(x, model, layers):

    features = {}
    for index, layer in model._modules.items():
```

```
        x = layer(x)
        if index in layers:
            features[layers[index]] = x

    return features
```

Next, the function should be called for both the **content** and **style** images:

```
content_features = features_extractor(content_img, \
                                      model, \
                                      relevant_layers)

style_features = features_extractor(style_img, model, \
                                    relevant_layers)
```

8. Calculate the gram matrix for the style features. Also, create the initial target image.

The following code snippet creates the gram matrix for each of the layers that was used to extract style features:

```
style_grams = {}
for i in style_features:
    layer = style_features[i]
    _, d1, d2, d3 = layer.shape
    features = layer.view(d1, d2 * d3)
    gram = torch.mm(features, features.t())
    style_grams[i] = gram
```

Next, the initial target image is created as a clone of the content image:

```
target_img = content_img.clone().requires_grad_(True)
```

If your machine has a GPU available, use the following code snippet instead:

```
target_img = content_img.clone().\
             requires_grad_(True).to(device)
```

9. Set the weights for different style layers, as well as the weights for the content and style losses:

```
style_weights = {'conv1_1': 1., 'conv2_1': 0.8, \
                 'conv3_1': 0.6, 'conv4_1': 0.4, \
                 'conv5_1': 0.2}


alpha = 1
beta = 1e5
```

10. Run the model for 500 iterations. Define the Adam optimization algorithm before starting to train the model, using **0.001** as the learning rate:

> **NOTE**
>
> To achieve the resulting target image shown in this book, the code was run for 5,000 iterations instead, which takes a very long time to run without a GPU. However, to appreciate the changes that start to happen in the output image, it is enough to run it for just 500 iterations, though you are encouraged to test different training times.

```
print_statement = 500
optimizer = torch.optim.Adam([target_img], lr=0.001)
iterations = 5000


for i in range(1, iterations+1):
    # Extract features for all relevant layers
    target_features = features_extractor(target_img, model, \
                                         relevant_layers)


    # Calculate the content loss
    content_loss = torch.mean((target_features['conv4_2'] \
                             - content_features['conv4_2'])**2)
```

```python
# Loop through all style layers
style_losses = 0
for layer in style_weights:

    # Create gram matrix for that layer
    target_feature = target_features[layer]
    _, d1, d2, d3 = target_feature.shape

    target_reshaped = target_feature.view(d1, d2 * d3)
    target_gram = torch.mm(target_reshaped, \
                            target_reshaped.t())
    style_gram = style_grams[layer]

    # Calculate style loss for that layer
    style_loss = style_weights[layer] * \
                torch.mean((target_gram - \
                            style_gram)**2)

    #Calculate style loss for all layers
    style_losses += style_loss / (d1 * d2 * d3)

# Calculate the total loss
total_loss = alpha * content_loss + beta * style_losses

# Perform back propagation
optimizer.zero_grad()
total_loss.backward()
optimizer.step()

# Print the target image
if i % print_statement == 0 or i == 1:
    print('Total loss: ', total_loss.item())
    plt.imshow(tensor2image(target_img))
    plt.show()
```

11. Plot the **content**, **style**, and **target** images to compare the results:

```
fig, (ax1, ax2, ax3) = plt.subplots(1, 3, figsize=(15, 5))
ax1.imshow(tensor2image(content_img))
ax2.imshow(tensor2image(target_img))
ax3.imshow(tensor2image(style_img))
plt.show()
```

The plots that are derived from this code snippet should be similar to the ones shown here:

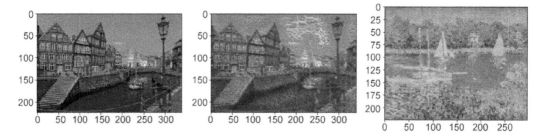

Figure 5.11: Output plots

NOTE

To view the high-quality color image, visit this book's GitHub repository at https://packt.live/2KcORcw.

To access the source code for this specific section, please refer to https://packt.live/2BZj91B.

This section does not currently have an online interactive example, and will need to be run locally.

To access the GPU version of this source code, please refer to https://packt.live/3eNfvqc. This version of the source code is not available as an online interactive example, and will need to be run locally with the GPU setup.

CHAPTER 6: ANALYZING THE SEQUENCE OF DATA WITH RNNS

ACTIVITY 6.01: USING A SIMPLE RNN FOR A TIME SERIES PREDICTION

SOLUTION

1. Import the required libraries, as follows:

```
import pandas as pd
import matplotlib.pyplot as plt
import torch
from torch import nn, optim
```

2. Load the dataset and then slice it so that it contains all the rows but only the columns from index 1 to 52:

```
data = pd.read_csv("Sales_Transactions_Dataset_Weekly.csv")
data = data.iloc[:,1:53]
data.head()
```

The output is as follows:

	W0	W1	W2	W3	W4	W5	W6	W7	W8	W9	...	W42	W43	W44	W45	W46	W47	W48	W49	W50	W51
0	11	12	10	8	13	12	14	21	6	14	...	4	7	8	10	12	3	7	6	5	10
1	7	6	3	2	7	1	6	3	3	3	...	2	4	5	1	1	4	5	1	6	0
2	7	11	8	9	10	8	7	13	12	6	...	6	14	5	5	7	8	14	8	8	7
3	12	8	13	5	9	6	9	13	13	11	...	9	10	3	4	6	8	14	8	7	8
4	8	5	13	11	6	7	9	14	9	9	...	7	11	7	12	6	6	5	11	8	9

Figure 6.26: Displaying dataset for columns from index 1 to 52

3. Plot the weekly sales transactions of five randomly chosen products from the entire dataset. Use a random seed of **0** when performing random sampling in order to achieve the same results as in the current activity:

```
plot_data = data.sample(5, random_state=0)
x = range(1,53)
plt.figure(figsize=(10,5))

for i,row in plot_data.iterrows():
    plt.plot(x,row)

plt.legend(plot_data.index)
```

```
plt.xlabel("Weeks")
plt.ylabel("Sales transactions per product")
plt.show()
```

The resulting plot should appear as follows:

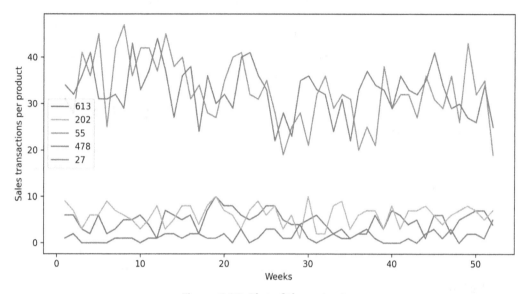

Figure 6.27: Plot of the output

4. Create the **inputs** and **targets** variables that will be fed to the network to create the model. These variables should be of the same shape and be converted into PyTorch tensors.

   The **inputs** variable should contain the data for all the products for all the weeks except the last week, since the idea of the model is to predict this final week.

   The **targets** variable should be one step ahead of the **inputs** variable; that is, the first value of the **targets** variable should be the second one of the inputs variable and so on until the last value of the **targets** variable (which should be the last week that was left outside of the **inputs** variable):

```
data_train = data.iloc[:,:-1]
inputs = torch.Tensor(data_train.values).unsqueeze(1)

targets = data_train.shift(-1, axis="columns", \
                    fill_value=data.iloc[:,-1])\
                    .astype(dtype = "float32")
targets = torch.Tensor(targets.values)
```

5. Create a class containing the architecture of the network. Note that the output size of the fully connected layer should be **1**:

```
class RNN(nn.Module):
    def __init__(self, input_size, hidden_size, num_layers):
        super().__init__()
        self.hidden_size = hidden_size
        self.rnn = nn.RNN(input_size, hidden_size, \
                          num_layers, batch_first=True)
        self.output = nn.Linear(hidden_size, 1)

    def forward(self, x, hidden):
        out, hidden = self.rnn(x, hidden)
        out = out.view(-1, self.hidden_size)
        out = self.output(out)

        return out, hidden
```

As in the previous activities, the class contains an **__init__** method, along with the network architecture, and a **forward** method that determines the flow of the information through the layers.

6. Instantiate the **class** function containing the model. Feed the input size, the number of neurons in each recurrent layer (**10**), and the number of recurrent layers (**1**):

```
model = RNN(data_train.shape[1], 10, 1)
model
```

Running the preceding code displays the following output:

```
RNN(
  (rnn): RNN(51, 10, batch_first=True)
  (output): Linear(in_features=10, out_features=1, bias=True)
)
```

7. Define a loss function, an optimization algorithm, and the number of epochs to train the network. Use the MSE loss function, the Adam optimizer, and 10,000 epochs to do this:

```
loss_function = nn.MSELoss()
optimizer = optim.Adam(model.parameters(), lr=0.001)
epochs = 10000
```

8. Use a **for** loop to perform the training process by going through all the epochs. In each epoch, a prediction must be made, along with the subsequent calculation of the loss function and the optimization of the parameters of the network. Save the loss of each of the epochs:

> **NOTE**
>
> Considering that no batches were used to go through the dataset, the **hidden** variable is not actually being passed from batch to batch (rather, the hidden state is used while each element of the sequence is being processed), but it was left here for clarity.

```
losses = []
for i in range(1, epochs+1):
    hidden = None
    pred, hidden = model(inputs, hidden)
    target = targets[:,-1].unsqueeze(1)

    loss = loss_function(targets, pred)
    optimizer.zero_grad()
    loss.backward()
    optimizer.step()
    losses.append(loss.item())

    if i%1000 == 0:
        print("epoch: ", i, "=... Loss function: ", losses[-1])
```

The output should look as follows:

```
epoch:  1000 ... Loss function:  58.48879623413086
epoch:  2000 ... Loss function:  24.934917449951172
epoch:  3000 ... Loss function:  13.247632026672363
epoch:  4000 ... Loss function:  9.884735107421875
epoch:  5000 ... Loss function:  8.778228759765625
epoch:  6000 ... Loss function:  8.025042533874512
epoch:  7000 ... Loss function:  7.622503757476807
epoch:  8000 ... Loss function:  7.4796295166015625
epoch:  9000 ... Loss function:  7.351718902587891
epoch:  10000 ... Loss function:  7.311776161193848
```

9. Plot the losses of all epochs, as follows:

```
x_range = range(len(losses))
plt.plot(x_range, losses)
plt.xlabel("epochs")
plt.ylabel("Loss function")
plt.show()
```

The resulting plot should appear as follows:

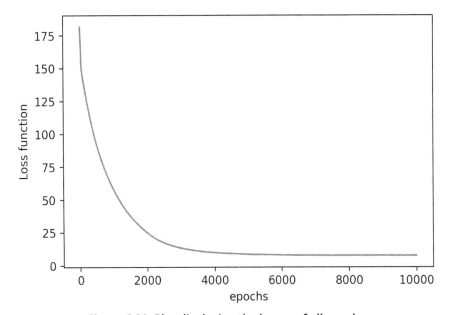

Figure 6.28: Plot displaying the losses of all epochs

10. Using a scatter plot, display the predictions that were obtained in the last epoch of the training process against the ground truth values (that is, the sales transactions of the last week):

```
x_range = range(len(data))
target = data.iloc[:,-1].values.reshape(len(data),1)

plt.figure(figsize=(15,5))
plt.scatter(x_range[:20], target[:20])
plt.scatter(x_range[:20], pred.detach().numpy()[:20])
plt.legend(["Ground truth", "Prediction"])
plt.xlabel("Product")
plt.ylabel("Sales Transactions")
plt.xticks(range(0, 20))
plt.show()
```

The final plot should be as follows:

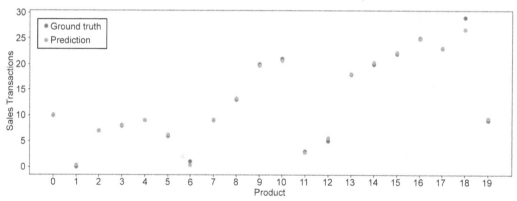

Figure 6.29: Scatter plot displaying predictions

NOTE

To access the source code for this specific section, please refer to https://packt.live/2BqDWvg.

You can also run this example online at https://packt.live/3ihPgKB.
You must execute the entire Notebook in order to get the desired result.

ACTIVITY 6.02: TEXT GENERATION WITH LSTM NETWORKS

SOLUTION

1. Import the required libraries, as follows:

```
import math
import numpy as np
import matplotlib.pyplot as plt
import torch
from torch import nn, optim
import torch.nn.functional as F
```

2. Open and read the text from *Alice in Wonderland* into the notebook. Print an extract of the first 50 characters and the total length of the text file:

```
with open('alice.txt', 'r', encoding='latin1') as f:
    data = f.read()

print("Extract: ", data[:50])
print("Length: ", len(data))
```

3. Create a variable containing a list of the unduplicated characters in your dataset. Then, create a dictionary that maps each character to an integer, where the characters will be the keys and the integers will be the values:

```
chars = list(set(data))
indexer = {char: index for (index, char) in enumerate(chars)}
```

The output should look as follows:

```
Extract:   ALICE was beginning to get very tired of sitting b
Length:   145178
```

4. Encode each letter of your dataset to its paired integer. Print the first 50 encoded characters and the total length of the encoded version of your dataset:

```
indexed_data = []
for c in data:
    indexed_data.append(indexer[c])

print("Indexed extract: ", indexed_data[:50])
print("Length: ", len(indexed_data))
```

The output is as follows:

```
Indexed extract:   [51, 52, 29, 38, 28, 25, 11, 59, 39, 25, 16, 53, 2,
1, 26, 26, 1, 26, 2, 25, 56, 60, 25, 2, 53, 56, 25, 23, 53, 7, 45,
25, 56, 1, 7, 53, 13, 25, 60, 14, 25, 39, 1, 56, 56, 1, 26, 2, 25,
16]
Length:   145178
```

5. Create a function that takes in a batch and encodes it as a one-hot matrix:

```
def index2onehot(batch):

    batch_flatten = batch.flatten()
    onehot_flat = np.zeros((batch.shape[0] \
                            * batch.shape[1],len(indexer)))
    onehot_flat[range(len(batch_flatten)), batch_flatten] = 1
    onehot = onehot_flat.reshape((batch.shape[0], \
                                  batch.shape[1], -1))

    return onehot
```

This function takes a two-dimensional matrix and flattens it. Next, it creates a zero-filled matrix of the shape of the flattened matrix and the length of the dictionary containing the alphabet (created in *Step 3*). Next, it fills the letter that corresponds to each character in the batch with ones. Finally, it reshapes the matrix so that it's three-dimensional.

6. Create a class that defines the architecture of the network. This class should contain an additional function that initializes the states of the LSTM layers:

```
class LSTM(nn.Module):
    def __init__(self, char_length, hidden_size, n_layers):
        super().__init__()
        self.hidden_size = hidden_size
        self.n_layers = n_layers
        self.lstm = nn.LSTM(char_length, hidden_size,\
                            n_layers, batch_first=True)
        self.output = nn.Linear(hidden_size, char_length)

    def forward(self, x, states):
        out, states = self.lstm(x, states)
        out = out.contiguous().view(-1, self.hidden_size)
```

```
        out = self.output(out)
        return out, states

    def init_states(self, batch_size):
        hidden = next(self.parameters())\
                    .data.new(self.n_layers, batch_size, \
                    self.hidden_size).zero_()
        cell = next(self.parameters())\
                .data.new(self.n_layers,batch_size, \
                self.hidden_size).zero_()
        states = (hidden, cell)
        return states
```

This class contains an **__init__** method where the architecture of the network is defined, a **forward** method to determine the flow of the data through the layers, and an **init_state** method to initialize the hidden and cell states with zeros.

7. Determine the number of batches to be created out of your dataset, bearing in mind that each batch should contain 100 sequences, each with a length of 50. Next, split the encoded data into 100 sequences:

```
# Number of sequences per batch
n_seq = 100
seq_length = 50
n_batches = math.floor(len(indexed_data) \
            / n_seq / seq_length)

total_length = n_seq * seq_length * n_batches
x = indexed_data[:total_length]
x = np.array(x).reshape((n_seq,-1))
```

8. Instantiate your model by using **256** as the number of hidden units for a total of two recurrent layers:

```
model = LSTM(len(chars), 256, 2)
model
```

Running the preceding code displays the following output:

```
LSTM(
  (lstm): LSTM(70, 256, num_layers=2, batch_first=True)
  (output): Linear(in_features=256, out_features=70, bias=True)
)
```

If your machine has a GPU available, make sure to allocate the model to the GPU, using the following code snippet instead:

```
model = LSTM(len(chars), 256, 2).to("cuda")
```

9. Define the loss function and the optimization algorithms. Use the Adam optimizer and the cross-entropy loss to do this. Train the network for **20** epochs:

```
loss_function = nn.CrossEntropyLoss()
optimizer = optim.Adam(model.parameters(), lr=0.001)
epochs = 20
```

If your machine has a GPU available, try running the training process for **500** epochs instead:

```
epochs = 500
```

10. In each epoch, the data must be divided into batches with a sequence length of 50. This means that each epoch will have 100 batches, each with a sequence of 50:

```
losses = []
for e in range(1, epochs+1):
    states = model.init_states(n_seq)
    batch_loss = []

    for b in range(0, x.shape[1], seq_length):
        x_batch = x[:,b:b+seq_length]

        if b == x.shape[1] - seq_length:
            y_batch = x[:,b+1:b+seq_length]
            y_batch = np.hstack((y_batch, indexer["."] \
                        * np.ones((y_batch.shape[0],1))))
        else:
            y_batch = x[:,b+1:b+seq_length+1]

        x_onehot = torch.Tensor(index2onehot(x_batch))
        y = torch.Tensor(y_batch).view(n_seq * seq_length)

        pred, states = model(x_onehot, states)
        loss = loss_function(pred, y.long())
        optimizer.zero_grad()
        loss.backward(retain_graph=True)
```

```
        optimizer.step()

        batch_loss.append(loss.item())

    losses.append(np.mean(batch_loss))

    if e%2 == 0:
        print("epoch: ", e, "... Loss function: ", losses[-1])
```

The output should look as follows:

```
epoch:  2 ... Loss function:  3.1667490992052802
epoch:  4 ... Loss function:  3.1473221943296235
epoch:  6 ... Loss function:  2.897721455014985
epoch:  8 ... Loss function:  2.567064647016854
epoch:  10 ... Loss function:  2.4197753791151375
epoch:  12 ... Loss function:  2.314083896834275
epoch:  14 ... Loss function:  2.2241266349266313
epoch:  16 ... Loss function:  2.1459227183769487
epoch:  18 ... Loss function:  2.0731402758894295
epoch:  20 ... Loss function:  2.0148646708192497
```

If your machine has a GPU available, the equivalent code snippet to train the network is as follows:

```
losses = []
for e in range(1, epochs+1):
    states = model.init_states(n_seq)
    batch_loss = []

    for b in range(0, x.shape[1], seq_length):
        x_batch = x[:,b:b+seq_length]

        if b == x.shape[1] - seq_length:
            y_batch = x[:,b+1:b+seq_length]
            y_batch = np.hstack((y_batch, indexer["."] \
```

```
                                    * np.ones((y_batch.shape[0],1))))
        else:
            y_batch = x[:,b+1:b+seq_length+1]

        x_onehot = torch.Tensor(index2onehot(x_batch))\
                .to("cuda")
        y = torch.Tensor(y_batch).view(n_seq * \
                            seq_length).to("cuda")

        pred, states = model(x_onehot, states)
        loss = loss_function(pred, y.long())
        optimizer.zero_grad()
        loss.backward(retain_graph=True)
        optimizer.step()

        batch_loss.append(loss.item())

    losses.append(np.mean(batch_loss))

    if e%50 == 0:
        print("epoch: ", e, "... Loss function: ", \
                losses[-1])
```

The result of running the training process for 500 epochs is as follows:

```
epoch:  50 ... Loss function:  1.5207843986050835
epoch:  100 ... Loss function:  1.006190665836992
epoch:  150 ... Loss function:  0.5197970939093622
epoch:  200 ... Loss function:  0.24446514968214364
epoch:  250 ... Loss function:  0.0640328845073437
epoch:  300 ... Loss function:  0.007852113484565553
epoch:  350 ... Loss function:  0.003644719101681278
epoch:  400 ... Loss function:  0.006955199634078248
epoch:  450 ... Loss function:  0.0030021724242973945
epoch:  500 ... Loss function:  0.0034294885518992768
```

As can be seen, by running the training process for more epochs, the loss function reaches lower values.

11. Plot the progress of the loss over time:

```
x_range = range(len(losses))
plt.plot(x_range, losses)
plt.xlabel("epochs")
plt.ylabel("Loss function")
plt.show()
```

The chart should appear as follows:

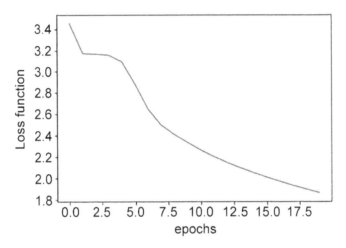

Figure 6.30: Chart displaying the progress of the loss function

As we can see, after 20 epochs, the loss function can still be reduced, which is why training for more epochs is strongly recommended in order to get a good result from the model.

12. Feed the following sentence **starter** into the trained model for it to complete the sentence: **"So she was considering in her own mind "**:

```
starter = "So she was considering in her own mind "
states = None
```

If your machine has a GPU available, allocate the model back to the CPU to perform predictions:

```
model    = model.to("cpu")
```

First, a **for** loop is used to feed the seed into the model so that the memory can be generated. Next, the predictions are performed, as can be seen in the following snippet:

```
for ch in starter:
    x = np.array([[indexer[ch]]])
    x = index2onehot(x)
    x = torch.Tensor(x)
    pred, states = model(x, states)


counter = 0
while starter[-1] != "." and counter < 100:
    counter += 1
    x = np.array([[indexer[starter[-1]]]])
    x = index2onehot(x)
    x = torch.Tensor(x)
    pred, states = model(x, states)
    pred = F.softmax(pred, dim=1)
    p, top = pred.topk(10)
    p = p.detach().numpy()[0]
    top = top.numpy()[0]
    index = np.random.choice(top, p=p/p.sum())

    starter += chars[index]

print(starter)
```

> **NOTE**
>
> To access the source code for this specific section, please refer to https://packt.live/2Bs6dRZ.
>
> This section does not currently have an online interactive example, and will need to be run locally.
>
> To access the GPU version of this source code, please refer to https://packt.live/3g9X6UI. This version of the source code is not available as an online interactive example, and will need to be run locally with the GPU setup.

ACTIVITY 6.03: PERFORMING NLP FOR SENTIMENT ANALYSIS

SOLUTION

1. Import the required libraries:

```
import pandas as pd
import numpy as np
import matplotlib.pyplot as plt
from string import punctuation
from sklearn.metrics import accuracy_score
import torch
from torch import nn, optim
import torch.nn.functional as F
```

2. Load the dataset containing a set of 1,000 product reviews from Amazon, which is paired with a label of **0** (for negative reviews) or **1** (for positive reviews). Separate the data into two variables – one containing the reviews and the other containing the labels:

```
data = pd.read_csv("amazon_cells_labelled.txt", sep="\t", \
                   header=None)
reviews = data.iloc[:,0].str.lower()
sentiment = data.iloc[:,1].values
```

3. Remove the punctuation from the reviews:

```
for i in punctuation:
    reviews = reviews.str.replace(i,"")
```

4. Create a variable containing the vocabulary of the entire set of reviews. Additionally, create a dictionary that maps each word to an integer, where the words will be the keys and the integers will be the values:

```
words = ' '.join(reviews)
words = words.split()
vocabulary = set(words)
indexer = {word: index for (index, word) \
           in enumerate(vocabulary)}
```

5. Encode the reviews data by replacing each word in a review with its paired integer:

```
indexed_reviews = []
for review in reviews:
    indexed_reviews.append([indexer[word] \
                            for word in review.split()])
```

6. Create a class containing the architecture of the network. Make sure that you include an embedding layer:

```
class LSTM(nn.Module):
    def __init__(self, vocab_size, embed_dim, \
                 hidden_size, n_layers):
        super().__init__()
        self.hidden_size = hidden_size
        self.embedding = nn.Embedding(vocab_size, embed_dim)
        self.lstm = nn.LSTM(embed_dim, hidden_size, \
                            n_layers, batch_first=True)
        self.output = nn.Linear(hidden_size, 1)

    def forward(self, x):
        out = self.embedding(x)
        out, _ = self.lstm(out)
        out = out.contiguous().view(-1, self.hidden_size)
        out = self.output(out)
        out = out[-1,0]
        out = torch.sigmoid(out).unsqueeze(0)

        return out
```

The class contains an **__init__** method, which defines the network architecture, and a **forward** method, which determines the way in which the data flows through the different layers.

7. Instantiate the model using 64 embedding dimensions and 128 neurons for three LSTM layers:

```
model = LSTM(len(vocabulary), 64, 128, 3)
model
```

Running the preceding code will display the following output:

```
LSTM(
    (embedding): Embedding(1905, 64)
    (lstm): LSTM(64, 128, num_layers=3, batch_first=True)
    (output): Linear(in_features=128, out_features=1, bias=True)
)
```

8. Define the loss function, an optimization algorithm, and the number of epochs to train for. For example, you can use the binary cross-entropy loss as the loss function, the Adam optimizer, and train for 10 epochs:

```
loss_function = nn.BCELoss()
optimizer = optim.Adam(model.parameters(), lr=0.001)
epochs = 10
```

9. Create a **for** loop that goes through the different epochs and through every single review individually. For each review, perform a prediction, calculate the loss function, and update the parameters of the network. Additionally, calculate the accuracy of the network on that training data:

```
losses = []
acc = []
for e in range(1, epochs+1):
    single_loss = []
    preds = []
    targets = []
    for i, r in enumerate(indexed_reviews):
        if len(r) <= 1:
            continue
```

```
x = torch.Tensor([r]).long()
y = torch.Tensor([sentiment[i]])

pred = model(x)
loss = loss_function(pred, y)
optimizer.zero_grad()
loss.backward()
optimizer.step()

final_pred = np.round(pred.detach().numpy())
preds.append(final_pred)
targets.append(y)
single_loss.append(loss.item())

losses.append(np.mean(single_loss))
accuracy = accuracy_score(targets,preds)
acc.append(accuracy)
if e%1 == 0:
    print("Epoch: ", e, "... Loss function: ", losses[-1], \
        "... Accuracy: ", acc[-1])
```

As in the previous activities, the training process consists of making a prediction, comparing it with the ground truth to calculate the loss function, and performing a backward pass to minimize the loss function.

10. Plot the progress of the loss and accuracy over time. The following code is used to plot the loss function:

```
x_range = range(len(losses))
plt.plot(x_range, losses)
plt.xlabel("epochs")
plt.ylabel("Loss function")
plt.show()
```

The plot should appear as follows:

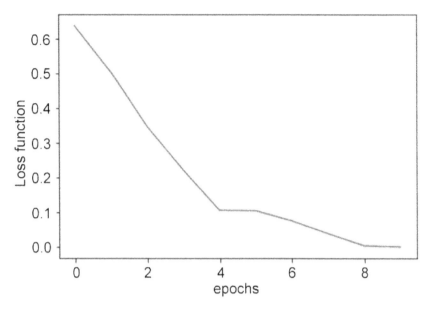

Figure 6.31: Plot displaying the progress of the loss function

The following code is used to plot the accuracy score:

```
x_range = range(len(acc))
plt.plot(x_range, acc)
plt.xlabel("epochs")
plt.ylabel("Accuracy score")
plt.show()
```

The plot should appear as follows:

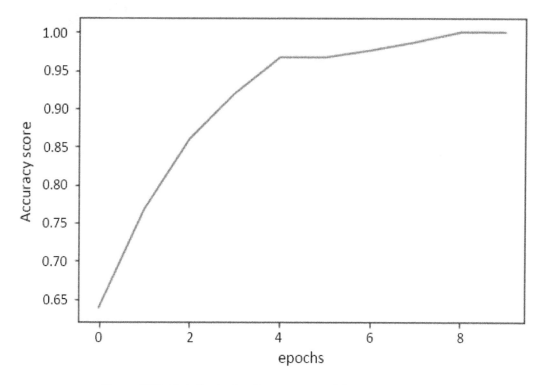

Figure 6.32: Plot displaying the progress of the accuracy score

NOTE

To access the source code for this specific section, please refer to
https://packt.live/2VyX0ON.

This section does not currently have an online interactive example, and will
need to be run locally.

INDEX

A

algorithm: 3, 18-19, 23, 28, 38-41, 50-51, 57, 68, 70, 74, 81, 96, 116, 162, 176-177, 179, 183, 205, 220, 229, 231
Anaconda: 81
arbitrary: 211
argmax: 89
arrays: 4, 8, 10
augmented: 145
autograd: 12-13, 17-18, 23

B

batchnorm: 66, 151
binary: 28, 31, 34, 37, 43, 78, 89, 231

C

chatbots: 2, 189-190
checkpoint: 106-107, 110, 113
citation: 64
classifier: 92-95, 106-107, 110, 150, 168
clustering: 89, 204
compiler: 108
concat: 83, 87-88
config: 109, 111
container: 13, 15, 65-66, 68, 91-92, 134, 137, 139
cosine: 34
counter: 222

covariance: 149
criterion: 95, 97

D

database: 161
dataframe: 22, 53, 58, 87-88, 198-199
dataloader: 141-142
derivative: 12

E

embedding: 227, 229, 231
energydata: 52
enumerate: 212, 216
epochs: 95-97, 99, 104-105, 143, 148, 152, 203, 205, 220-221, 224-225, 229, 231-232

F

figsize: 86, 181
filename: 107
filter: 44-47, 121, 127-133, 135, 137, 143
flattening: 143
fontsize: 86
framework: 5, 65, 108, 142
frontend: 5, 23

G

GitHub: 15, 21, 52, 64, 67, 77, 142-143, 149, 153, 159, 164-165, 178, 181, 183, 223-224
gradient: 12, 17, 39-40, 233
grayscale: 122, 140, 146

H

hyperbolic: 34
hyperlink: 21
hypertext: 109

I

imagenet: 161
imputation: 51
imshow: 166, 175, 181
imsize: 164
indexer: 212, 216-217, 220, 222
integer: 224, 228, 231
intercept: 17
interface: 11, 108
invariance: 57, 145
isnull: 53, 83
iteration: 17-18, 22, 36-39, 68-70, 104, 141, 177-179

J

Jupyter: 9, 19, 52, 56, 58, 60, 67, 70, 81, 93, 104, 111, 113-114, 164, 198

L

linearity: 31-32, 66, 132, 196
loader: 141, 164-165, 182

M

matplotlib: 9, 19, 86, 93, 98, 162-164, 181, 198
maxpool: 136-139, 151
metric: 75, 77-78, 86, 96, 101, 103
minima: 38-39
module: 13-15, 17-18, 23, 61, 65-67, 91-93, 98-99, 108, 114, 116, 133-134, 136-139, 151, 202, 219, 228-229
multiclass: 31

N

nonetype: 13

O

one-hot: 215-218, 221, 224, 228
operator: 190
ordinal: 88
outlier: 51, 56, 103

P

pandas: 21-22, 52, 61, 64, 81-82, 87-88, 93, 98, 198
perceptron: 26, 28-30, 48, 71
player: 225
pooling: 45-48, 126, 132, 134-137, 139, 143, 154, 168-170, 184
pydata: 22
pyplot: 19, 86, 98, 162, 164
Python: 5, 8-9, 23, 52-53, 89, 93, 107-109, 111, 113, 163-164, 198
pythonic: 108
PyTorch: 1-2, 5-13, 15, 17, 19, 21, 23, 26, 65, 67, 73, 89, 91, 93-94, 105-108, 113, 116, 119, 121, 126, 129, 133, 136, 138, 140-142, 145, 151, 154, 157, 160, 162, 164, 168-169, 184, 198, 202, 205, 211, 221, 227-228, 233

R

randint: 8, 19, 198
regression: 23, 25-26, 37, 51, 68-71, 74, 89, 91, 106, 188
reloader: 111-112
repository: 15, 21, 52, 64, 67, 77, 142-143, 149, 153, 164-165, 178, 181, 183, 223-224, 230
rmsprop: 17

S

sampler: 141-142
sequential: 13-16, 65-66, 68, 91-92, 134, 137, 139, 185, 188
sigmoid: 14-16, 33-34, 66, 89, 209-210
sklearn: 61, 98
snippet: 6, 8, 16, 18, 55, 58, 69, 82, 87, 92, 95-96, 98, 133-134, 136, 138, 140-141, 146, 162, 165-166, 169-171, 175, 212, 214, 216, 220, 222-223, 228
softmax: 34, 66, 89, 92-95, 99, 126, 138-139, 143, 151, 222
stochastic: 17, 40
subfield: 226
subgroups: 43
subpackage: 160, 168-169, 184
subplots: 86, 181
subsection: 44, 46
subset: 2, 23, 90

T

tensor: 4, 6-10, 12-13,

19, 66-67, 97, 110,
114, 163, 166, 175,
181-182, 220-222

TensorFlow: 11
toolkit: 10
topology: 90
transpose: 83

V

variance: 102-105,
116, 149
vector: 121, 137

X

xticks: 86